D1164026

USING WAVES AND VHDL FOR EFFECTIVE DESIGN AND TESTING

Using WAVES and VHDL for Effective Design and Testing

A *practical* and *useful* tutorial and application guide for the Waveform and Vector Exchange Specification (WAVES)

James P. Hanna
USAF Rome Laboratory, Rome, NY

Robert G. Hillman
USAF Rome Laboratory, Rome, NY

Herb L. Hirsch
MTL Systems, Inc., Dayton, OH

Tim H. Noh
MTL Systems, Inc., Dayton, OH

Dr. Ranga R. Vemuri
University of Cincinnati, Cincinnati, OH

Includes ready-to-use examples, tools and templates

USING WAVES AND VHDL FOR EFFECTIVE DESIGN AND TESTING

James P. Hanna
USAF Rome Laboratory

Robert G. Hillman
USAF Rome Laboratory

Herb L. Hirsch
MTL Systems, Inc.

Tim H. Noh
MTL Systems, Inc.

Ranga R. Vemuri
University of Cincinnati

KLUWER ACADEMIC PUBLISHERS
Boston / Dordrecht / London

Distributors for North America:
Kluwer Academic Publishers
101 Philip Drive
Assinippi Park
Norwell, Massachusetts 02061 USA

Distributors for all other countries:
Kluwer Academic Publishers Group
Distribution Centre
Post Office Box 322
3300 AH Dordrecht, THE NETHERLANDS

Library of Congress Cataloging-in-Publication Data

A C.I.P. Catalogue record for this book is available from the Library of Congress.

Printed on acid-free paper.

Printed in the United States of America

Using WAVES and VHDL for Effective Design and Testing

Dedications

From James P. Hanna

... to my family for their support and guidance

From Robert G. Hillman

... to my wife, Linda, for her enduring support and encouragement through out the years

From Herb L. Hirsch

... to my wife, Susan Diane, for her support and encouragement of my techno-dabbling

From Tim H. Noh

... to my wife, Misuk, and my children, Ada, Paula, Michael, and Jason

From Ranga R. Vemuri

... to my grandfather Sri V. L. Narasimham for his early efforts in guiding me into the right path of life

Using WAVES and VHDL for Effective Design and Testing

Table of Contents

List of Figures

List of Tables

Preface

The proliferation and growth of Electronic Design Automation (EDA) has spawned many diverse and interesting technologies, of which one of the most prominent is the VHSIC Hardware Description Language, or *VHDL*. In fact, the significance of VHDL is of such magnitude that some might argue that VHDL spawned EDA, and not the other way around. This is obviously a cart-and-horse issue and we will not linger upon it here. In any case, the fact remains that VHDL is a significant foundation technology in EDA. VHDL permits designers of digital modules, components, systems, and even networks to describe their designs both structurally and behaviorally, and to simulate them to investigate their performance, prior to actually implementing them in hardware. This simulate-before-you-build capability is VHDL's forte, and numerous EDA tool suites have been built upon VHDL, providing the designer with graphical interfaces, libraries, post-simulation analyses, and other supporting functions to facilitate the use of VHDL for myriad design projects.

Having gained the ability to simulate designs, once coded in VHDL, designers were naturally confronted with the issue of *testing* these designs. Here, the requirement to insert particular digital waveforms, often termed *test vectors* or *patterns*, and to subsequently assess the correctness of the response from some digital entity, was not explicitly addressed in VHDL. Designers are creative individuals, however (otherwise they would not be designers), and many turned to *de-facto* methods for generating test vectors and assessing responses. Since VHDL is a high-level programming language, test pattern generators could be coded, as could response comparison algorithms, and many designers used this approach. In certain cases, the EDA tool suites contained some test pattern generation and results analysis capability as well, and designers owning such tools could simply use them. In an isolated design environment, these methods were satisfactory.

However, in a distributed design environment, or even in an isolated one where the design was subject to review or scrutiny by another (than the designers) organization, these *de-facto* methods of testing and evaluating results proved faulty. The reason was simply a lack of standardization. When organization A designed a circuit and tested it with their self-developed test tools it had a certain behavior. Then when it was delivered to organization B, and they tested it using their test tools, the behavior was different. What was the problem? Was the circuit, A's tools, or B's tools faulty? The only way to resolve this was for both organizations to agree on a test apparatus, validate its correctness, and use it consistently. Then they had to do it all over again for the next circuit which was beyond the scope of the current test tools. Then they had to put the evolving test apparatus under some kind of configuration control to manage it. While VHDL was an IEEE standard language, and consistency among myriad designers was fairly well guaranteed, no such standard existed for test waveform generation and assessment. Hence, the value of standardization in the design language was being negated by the lack of such a standard for testing.

The Waveform and Vector Exchange Specification, or *WAVES*, was conceived and designed to solve this testing problem - and it has. Being both a subset of VHDL itself, as well as an IEEE standard, it guarantees both conformity among multiple applications and easy integration with VHDL units under test (UUTs). However, it has historically suffered from one problem - designers have *perceived* that it is easier to write their own waveform generators and comparison algorithms than to use WAVES. Our purpose in writing this book is to destroy that myth once and for all. Using WAVES is not difficult at all, once we understand its fundamental structure and organization. Actually, designers choosing to write their own waveform generators and evaluators will wind up constructing the same kinds of modules they would for WAVES. However, by doing it in WAVES, these modules can be used by anyone wishing to replicate the testing, and this will guarantee a working, useful testbench (WAVES' term for the combined waveform generator, UUT, and results comparator). Furthermore, these testbenches can be used to program hardware testers as well, thus guaranteeing that the simulation testing and the hardware testing, upon implementing the circuit, will be compatible. No apples and oranges here -- just sound, standardized, coordinated testing of the simulatable VHDL and the hardware rendition.

So if WAVES is so useful, and easy to use, then why the perception problem? In our esteem, probably because most people initially attempt to use WAVES when they have a VHDL module to test, and try to simply "hook it up and make it run." This approach, coupled with the lack of a straightforward tutorial written from the designers' point of view, has been the source of designers' difficulties. Our goal in this book is to provide this needed tutorial, along with some advanced topics, to render WAVES' use no more difficult than any other design task in the VHDL-EDA world.

Our approach to this is simple -- start with the waveform and work toward the testbench implementation. All digital designers understand waveforms -- that is our medium. Hence, after providing some initial history and background for WAVES and VHDL, we take a digital waveform -- a series of events and logic level transitions to be applied to a UUT -- and dissect it into what we term "building-block" components. Then we describe how these building block components can be configured, assembled into a waveform representation, applied to a VHDL UUT, and compared with the outputs from this UUT. That is all there is to WAVES implementation and use. In the course of this sequence, we explicitly describe the WAVES elements, and provide practical examples, tools, and templates for the designer's use and understanding. Then, in the later chapters, once the basic composition and use of the WAVES testbench is thoroughly understood, we provide insights into some advanced topics such as added flexibility in use, complex timing applications, integration with testers, and more complex manipulation of the waveform injection and response evaluation including serial data exchange and boundary-scanning a device.

In total, we believe this text will serve many purposes. For the WAVES beginner, its tutorial will make the application of WAVES in typical, standard usage straightforward and convenient. For the more advanced user, the advanced topics will provide insight into the nuances of these useful capabilities. For all users, the tools, templates and examples, given in the chapters as well as on the companion disk, will provide a practical starting foundation for using WAVES and VHDL.

Before proceeding into the text, we would like to leave our readers with these thoughts: WAVES is not *difficult*, WAVES is *structured*, and anything structured becomes simple once we are familiar with the structure. Since we are already familiar with waveform structures, all we really need to understand is how WAVES represents the waveforms we already understand. It's really as simple as that, as we will now demonstrate in this text.

The Authors
August, 1996

Acknowledgments

In preparing this text, the authors received substantial support from many sources, which we feel greatly enhanced the quality and completeness of this work. To the following individuals and their parent organizations, we wish to express our sincere thanks and appreciation:

Steve Drager, Chris Flynn, and Jim Nagy of Rome Laboratory, for their assistance in developing the VHDL models for the manuscript, and for their review and critique of the manuscript.

Mark Pronobis of Rome Laboratory, for inspiring the development of the WAVES toolset and his assistance throughout the tool evaluation and debug process.

Mike Doll of the University of Cincinnati, for preparing the calculator example used in Chapter 10 and for reviewing the drafts of Chapters 10 through 12.

Dr. Praveen Chawla of MTL Systems, Inc., for his review and critique of the code examples and templates, and assistance in designing the indices.

Karen Winters of MTL Systems, Inc., for her efforts in preparing, editing, and assembling the manuscript.

Jonathan Nocjar of MTL Systems, Inc., for producing the CD-ROM master.

The management of MTL Systems, Inc., for providing the resources to produce the manuscript and CD-ROM.

Carl Harris of Kluwer Academic Publishers for exercising both patience and persistence as we developed the manuscript.

The preparation of the material in this book was sponsored in part by the United States Air Force Materiel Command.

CHAPTER 1. INTRODUCTION

The purpose of this WAVES/VHDL guidelines text, and how to use it

In this chapter, we introduce our text, in the context of its purpose and how to use it. We begin with an overview of **electronic design automation (EDA)** in general, and how this field is moving toward a tighter integration between the development and testing processes. Next, we discuss the role of **VHDL, the VHSIC Hardware Description Language**, and how it is proliferating and becoming the foundation for many design automation tools. Then, we briefly overview **WAVES, the Waveform and Vector Exchange Specification**, relating how it is intended to interface with VHDL to provide a robust and standardized test vector management paradigm. Finally, we overview our text, both in terms of its contents and how any user, from novice to experienced practitioner, will be able to use this material to support their particular design and testing needs.

1.1 Electronic Design Automation (EDA) and Testing

The goal of EDA is to create environments in which engineers may develop their concepts into implementations with effective, automated support tools. In the early days of EDA, this capability was basically provided by tools to support the various stages of the development process, as we illustrate in Figure 1-1. We had tools to help us with our concepts, specification-writing tools to ensure we produced good and complete specifications, design tools to help us create behavioral and structural designs in various stages, and implementation tools to permit us to create our designs in hardware. For example, we might design a digital subsystem using a **Computer-Aided Engineering (CAE)** or **Computer-Aided Design (CAD)** system, and then program it into a **field-programmable gate array (FPGA)** using an FPGA partitioning and layout tool. In this stage of EDA evolution, we were beginning to see the value of computer-aided assistance, but we were also beginning to get buried in myriad, disparate tools, with accompanying compatibility and redundancy problems.

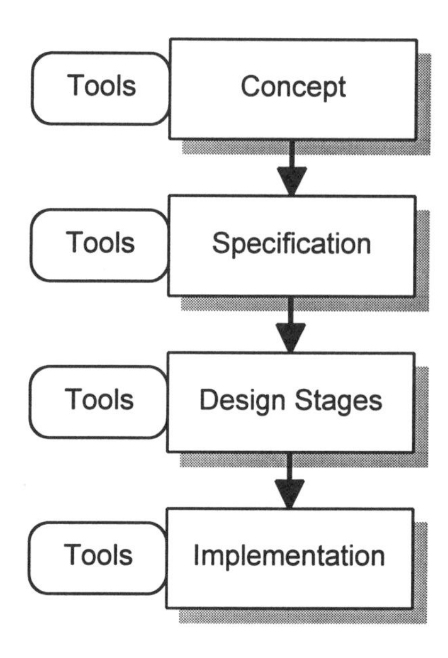

Figure 1-1. A nominal, multi-stage development process

From that beginning, we have progressed to the point where the EDA community at large, including both users and developers of the tools, are interested in more unified environments. Here, the notion is that the tools used at the various stages in the development process need to be able to *complement each other*, and to *communicate with one another efficiently* using effective file exchange capabilities. Furthermore, the idea of capturing all the tool support needed for an EDA development into a unified support environment is now becoming a reality. This reality is evidenced by some of the EDA suites we now see emerging, wherein several tool functions are integrated under a common **graphical user interface (GUI)**, with supporting file exchange and libraries to enable all tool functions to operate effectively and synergistically. This concept, which we illustrate in Figure 1-2, is the true future of EDA.

Within this unified development concept, a desired feature is the ability to evaluate our concepts, specifications, designs, and implementations at the various stages. Going one step further, we would like to have the ability to *simulate* our concepts at these stages, under a variety of static and dynamic stimulus conditions, and to obtain results which assist us in our decision-making, at all levels of the development process. Furthermore, we would like a tight integration among the various simulations, their test stimuli, and the actual testing we will apply to the hardware end-product or brassboard configurations. Such an integration permits coordinated testing and results analysis throughout the development process.

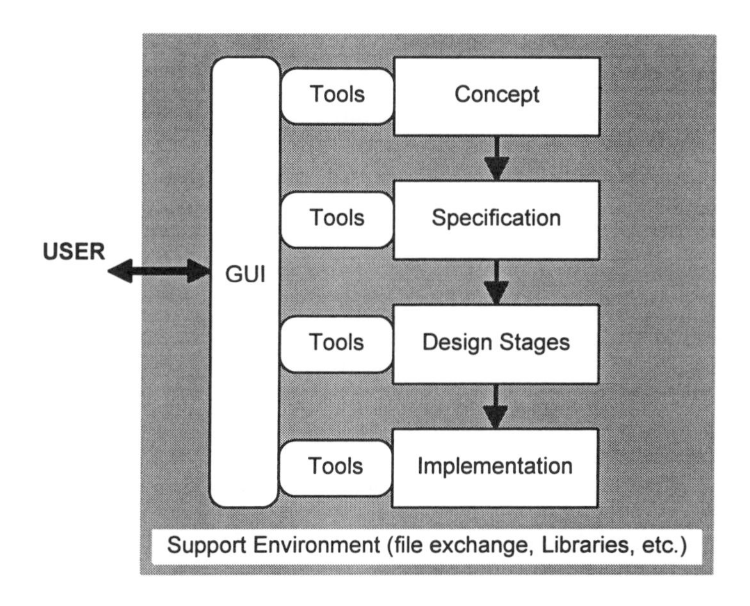

Figure 1-2. An example of a unified, multi-tool EDA suite

As this notion of unified support environments for the EDA development process became popular, we realized that some kind of executable, high-level description language would be essential to effectively unify the tool support environment and provide the means for simulation. Fortunately, that language already existed, and was ready to assume its proper role in this EDA unification process.

1.2 The VHSIC Hardware Description Language (VHDL)

VHDL was originally conceived as a system specification language. It was developed to allow digital designers to describe their logic designs both behaviorally and structurally, and then to simulate them by executing these VHDL descriptions, once they were analyzed and compiled. However, VHDL also had two important properties which made it attractive for other uses, at other levels and stages of the development process. First, it was hierarchical, so objects could have sub-objects, be part of super-objects and so forth. Hence, we could describe a system at whatever level of abstraction we chose, from system-level to gate-level, and our description would simply become more detailed as we added the definitions of lower-level pieces. Second, VHDL incorporated the concept of time, so time relationships could be captured and recorded, without the bother of writing variables to track timing nuances. These would prove to be valuable features to the EDA process.

EDA tool developers soon realized that if they made VHDL the underlying engine beneath their tool suites, a powerful and unifying design, simulation, and analysis capability could be obtained. Basically, as we show in Figure 1-3, we would be able to simulate our design at any level, from concept to implementation. We could model our concepts as highly-abstracted notions, provide as much behavioral definition we had at this stage, and simulate them in VHDL to get an early idea of performance. We could detail these notions into specifications, simulate them, and determine continuity, completeness and again, behavior to the level of definition we could provide. We could model our designs at every stage of refinement, and get equally refined performance simulation results as our VHDL definition became more elaborated along with the design. Finally, we could capture the final stages in a technology we call *synthesis*, to automatically produce the hardware renditions of our designs.

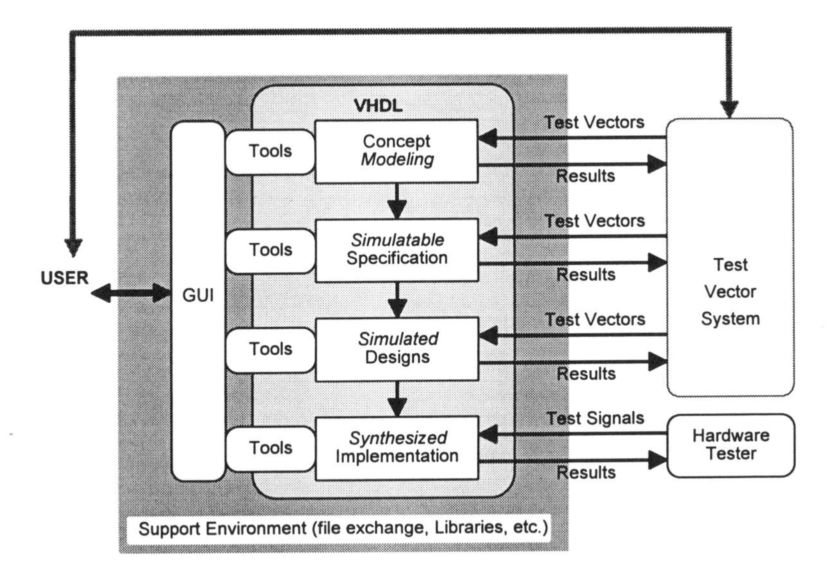

Figure 1-3. The VHDL foundation in the EDA tool suite concept

The majority of today's EDA suites are indeed following this paradigm, with performance modeling tools, simulatable specifications, design tools, and synthesis systems, all based upon VHDL. However, this unification of the EDA suite around VHDL has also precipitated a requirement, as Figure 1-3 illustrates - the need to stimulate our simulations at the various stages of development and to collect results. Basically, we need test vector generation and results collection and comparison for the simulated development descriptions at all stages, exactly as we have always required signal generation and data collection and comparison capability for the hardware end-product of such a process. WAVES was conceived to satisfy this need.

1.3 WAVES with VHDL

WAVES was designed to be the unified testing and results collection system to complement the unified development systems based upon VHDL. Its purpose is to provide the means to define test stimuli, in the form of digital waveforms, or test vectors, to define the results to be collected, and to manage the insertion of the stimuli and the collection and comparison of the results as the VHDL description is simulated. It is also designed for compatibility with hardware testers, such that the same test stimuli and collection paradigm may be automatically communicated to hardware test systems, to ensure identical testing of the hardware and the pre-implementation simulation. Basically, it is designed to be completely complementary to, and synergistic with, VHDL simulation and actual hardware testers.

This complementary and synergistic characteristic exists in WAVES for a simple reason - *WAVES is a proper subset of VHDL*. More precisely, as we illustrate in Figure 1-4, the testing and collection entity of WAVES, called the ***WAVES testbench***, is written in VHDL, attached to the VHDL description, it is analyzed, compiled, and executed along with the rest of the VHDL under simulation. Hence, all the utility and unification provided by VHDL complement WAVES. Obviously, it is a powerful tool.

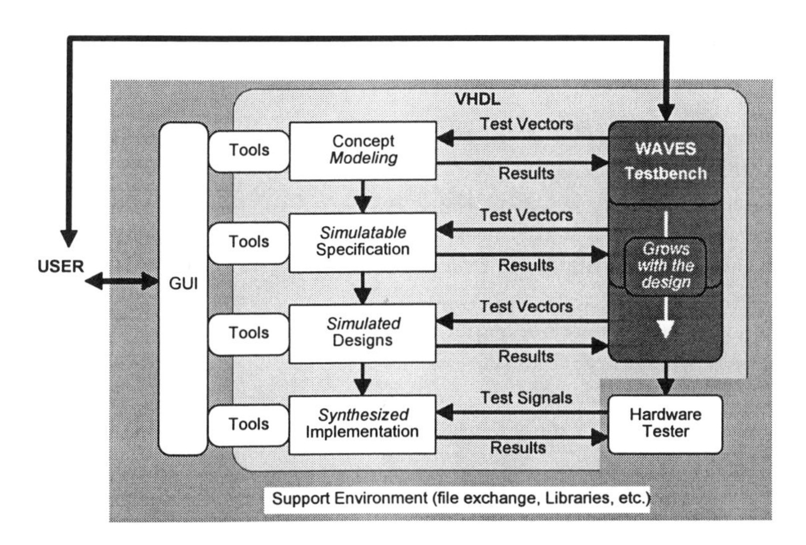

Figure 1-4. WAVES as the test system in the EDA tool suite concept

Although admittedly powerful, in order for WAVES to be useful as well, the **task of constructing and attaching the WAVES testbench needs to be an easy one**. It is very easy for a designer to simply write a test vector to assess a particular aspect of some design at some stage of development. As the design becomes more elaborate, it is also a simple matter for that designer to add more and more hand-crafted test vectors. However, doing so will not ensure rigorous testing of the design stages, or compatibility with hardware testing at the end. By following WAVES, a designer will obtain properly-scoped and standardized tests, and the WAVES testbenches so created can be shared among others in a design team, and with customers, for review. Basically, the standardization aspect ensures accurate communication of the testing and results among all who require this information, just as VHDL ensures such communication for the design itself. In other words, if we accept the benefits of standardized and structured design, we also accept those for the accompanying testing. Furthermore, an additional benefit of WAVES standardization lies in the aspect of test vector exchange. Basically, all parties using the standard can exchange their testbenches and datasets, to ensure common testing among developers, users, or evaluators of the designs and implementations.

However, as we pointed out earlier, if designers are to use WAVES, it must become a comfortable element of their working environment. For example, structured and standardized design methodologies only became popular when their support tools became easy to use. Similarly, WAVES will realize its full potential only when it becomes easy to use, and that is precisely the purpose of this text. Our goal is to provide the instruction, examples, and understanding of WAVES to render it a valuable and easily-applied design support tool.

1.4 Overview of the Text

Our text is designed to provide useful and necessary information for a variety of readers. As such, we have included what we consider to be a user-friendly indexing paradigm. This includes two indexes, one by topic and one by application. The reader who knows a particular topic of interest (such as Testbench, Pin, and so forth) can look it up in the Topic Index. The reader who wants to perform a certain task (create a testbench, assign a pin, and so forth) will be able to look up these topics under the action words (create, assign, save, modify, and so forth) in the Application Index. In this manner we have sought to satisfy the needs of all readers.

We have organized the contents of the text into three portions: An *overview and historical perspective* (Chapters 1 and 2), a basic tutorial (Chapters 3-7), and dissertations of *advanced topics* (Chapters 8-12). Throughout these chapters we have included practical and useful example cases for the effective application of WAVES and VHDL. We now preview the contents of these three portions of the text.

For the novice user of WAVES and VHDL, we provide the *overview and historical perspective* and *basic tutorial* portions. Here, the reader can gain insights into the motivation behind, and development of, WAVES, and also follow a step-by-step tutorial to gain the skills necessary to construct, implement, and use the WAVES and VHDL capability. Our "running example" in the tutorial portion, which accompanies the discussions, illustrates the evolving concepts and techniques as they are presented.

We began the *overview and historical perspective* with this chapter, in which we have already described the status of EDA in general, the impact of VHDL, and the utility of WAVES. In Chapter 2, which concludes this portion, we describe the history and evolution of WAVES, to provide the reader with useful insight into its application as we proceed through the remainder of the text.

In the *basic tutorial* we offer step-by-step, building-block instruction for developing and applying WAVES and VHDL. First, in Chapter 3, we lay the foundation through an overview of the waveform terminology and conventions we will use throughout the text. Next we apply the foundation waveform concepts to constructing the particular elements of WAVES, in Chapter 4, by describing this construction process in detail and giving the reader an appreciation of how to implement the waveform concepts in the WAVES constructs. Once we have the constructs, the next step is to put them together. Hence, in Chapter 5 we take the WAVES elements we discussed in Chapter 4, and describe how to assemble them into a WAVES dataset. Then, in Chapter 6, we describe how to put everything together, to simulate the combined WAVES and VHDL and obtain the results we need to support our design and testing endeavors. Finally, in Chapter 7 we describe the details of an important entity, the external file, which provides the information necessary to orchestrate the WAVES building-block method of assembling waveforms.

For the intermediate user of WAVES, or the novice who has progressed through the tutorial, we provide the *davanced topics* portion, describing several complexity-expanding aspects and showing some useful examples. We begin this portion with Chapter 8, in which we introduce the issues relating to more complex, realistic timing relationships which WAVES permits us to handle. Next, since we often wish to exchange some known or existing waveform in a hardware system or tester with our WAVES and VHDL simulation, in Chapter 9 we present the means to accomplish this waveform exchange, to and from automatic test equipment (ATE). At this point, having thoroughly described and demonstrated WAVES Level 1, we introduce and explain WAVES Level 2 in Chapter 10. Here we also discuss the relationships among these levels, to provide the reader with a practical appreciation of where and how each level can be most effectively applied to a design or testing problem. Our discussions also include some other added-complexity issues germane to effective use of WAVES at any level. Then, in Chapter 11, we discuss the application of waveform data exchange by which the waveform generation methods and the unit under test may interact with each other in more complex ways than a simple stimulus-response paradigm. Finally, in chapter 12 we look into serial waveform data exchange through boundary-scanning techniques.

In ***summary of Chapter 1***, we have overviewed design automation, described the synergism between WAVES and VHDL in effective design automation efforts, and provided guidance for effectively using this text. Next, before beginning our tutorial portion, we provide a brief history of WAVES, to acquaint the reader with its background, evolution, and capability to improve the effectiveness of design automation work.

CHAPTER 2. THE HISTORY AND BACKGROUND OF WAVES

The need to simplify

In this chapter, we summarize the history and evolution of WAVES. Our intent here is not to bog down the flow of this text with an inordinately elaborated historical dissertation. Rather, we feel that a notion of the history and development of WAVES will provide useful insight into its application, as we proceed through the remainder of the text. We begin with a brief description of the history of WAVES, concentrating upon the motivation behind its creation and its relationship to VHDL. Next, we overview the concept of integrating WAVES and VHDL for effective design automation practice, highlighting the needs for complexity management and for tool and library support to achieve this effectiveness. Finally, we summarize the use of certain key library components for integrating WAVES and VHDL, setting a conceptual reference for the more detailed technical elaborations of the later chapters. In aggregate, we expect this material to lay the proper foundation for the remainder of our text.

2.1 WAVES History

Our historical treatment of WAVES is designed to give the reader a perspective on the motivation behind the conception and development of WAVES. Here, we look at two particular aspects: (1) the need for WAVES in general, and (2) the rationale for its close ties to the VHSIC Hardware Description Language (VHDL). From this perspective, the reader will gain some valuable insights to carry forward into the later, more technically-explicit chapters, where questions such as, "Why did they do it that way?" often arise. The answers to many such questions lie within this history.

The development of WAVES resulted from innovators and researchers in electronic design automation realizing that a standard for effective test waveform exchange would be a critical and necessary asset to the evolution, proliferation and effectiveness of design automation in general. The original notion of such a standard began at the United States Air Force Rome Laboratory, and the early development work was supported by a variety of government and commercial organizations. Finally, WAVES became an IEEE Standard (IEEE STD 1029.1) in 1991, ready to assume its role in the electronic design automation world.

The name "WAVES" was carefully chosen. The words "waveform" and "vector" indicate that WAVES may represent simulator event trace data, as well as the highly-structured test vectors typical of automated test equipment. The word "exchange" means that WAVES is meant for the exchange of information among vendors and users as well as between design and test environments. WAVES was not, however, designed to replace the particular stimulus or response formats used within a given environment.

Because WAVES is an *exchange specification*, all facets of the stimulus and response data must be captured. Nothing may be left to the user's imagination, since everyone has slightly different interpretations about what constitutes expected data, and what is "obviously right" to one engineer is likely to be "obviously wrong" to another. Hence, when we are exchanging information among environments, assumptions are dangerous and often incorrect. Therefore, WAVES data stands alone and does not require anything in common between the sender and the receiver, other than an adherence to the WAVES specification.

Although WAVES has strong and explicit ties to the VHSIC Hardware Description Language (VHDL), its utility and application are not constrained to VHDL practitioners. WAVES is a subset of IEEE STD 1076-1993, the VHDL standard. Hence, anyone with an understanding of VHDL will have an easy time understanding the syntax and semantics of WAVES. However, a complete knowledge of VHDL is certainly not a prerequisite for understanding WAVES. WAVES includes only the *sequential* portion of VHDL, which means that WAVES is essentially an algorithmic programming language. Anyone familiar with a modern programming language, such as Ada or C++, will have no trouble understanding WAVES source code.

VHDL was chosen as the basis for WAVES because VHDL is so important in the design phase of electronic components and modules. First of all, VHDL is a standard representation for modeling and simulating digital circuits. Also, extensive effort has gone into the design of VHDL, so there was no reason to repeat that work by designing a totally new language for WAVES. Finally, many design environments are available for VHDL, and these may be used, without change, on a WAVES dataset. These aspects of standardization, significant previous development, and available environments, rendered VHDL an ideal host for WAVES.

In ***summary of our WAVES history***, we note the motivations for both the development of WAVES and its ties to VHDL were based upon the goal to improve design automation effectiveness. Next, we will look more closely at how WAVES and VHDL can be combined in modeling and simulation applications for effective design automation practice.

2.2 WAVES-VHDL Integration for Modeling and Simulation

The fundamental notion of using WAVES and VHDL together is implicit within their common VHDL foundation. However, the actual implementation of an effective WAVES and VHDL *integrated* application is embedded within an entity called the ***WAVES testbench***. We will be covering the nuances of this testbench in detail in the later chapters, but the point here is that we *need to create the testbench* to manage the insertion of test waveforms and vectors into, and the examination of results data from, the VHDL design description we wish to test. As Figure 2.1 illustrates, the testbench is simply the entity which manages the test vector exchange. In fact, it is more appropriately called a WAVES-VHDL testbench, as it instantiates both the WAVES dataset and the VHDL model to be tested. The testbench needs to be properly created and attached to the VHDL design, so that they may be compiled and simulated together.

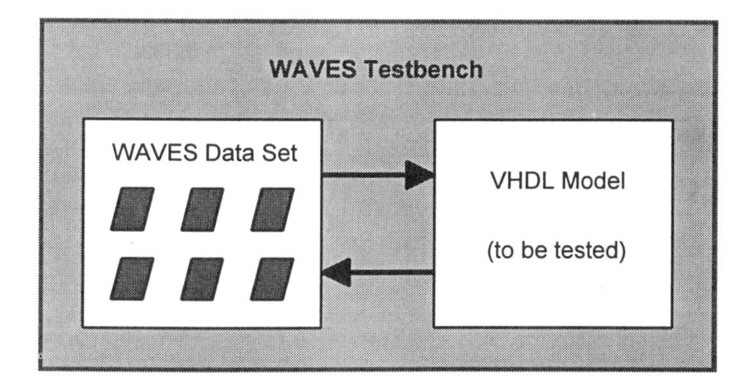

Figure 2-1. The WAVES testbench including a VHDL Model

The WAVES Standard provides all the framework and specification information required for creating this WAVES testbench. However, the acceptance of a standard as a common practice is often slow, and requires an iterative process involving user demand, support tooling, and user acceptability. Tool vendors will only invest in a tool development when it is apparent that there is a profitable market available. Users, on the other hand, require tool support as well as an understanding of the standard and the benefits of using that standard. The main ingredient in the adoption of a standard as a common practice is the availability of information for the use and application of the standard. In this section, we focus upon the work performed to enhance the integration of WAVES and VHDL for model simulation and verification.

The fundamental complexity of WAVES lies within its library and package management paradigm. In VHDL, libraries and their integral packages provide the means for sharing information among all entities which require access to that information. For WAVES, this package organization is shown in Figure 2-2, which basically illustrates that the packages comprising a user-developed WAVES dataset (User-Developed WAVES), the WAVES testbench (Testbench) and the VHDL model to be tested (VHDL Model), *all* need to come together for subsequent simulation. We will cover the specifics of constructing or using the entities shown within the figure, in detail, in the later chapters. Here, our point is simply the complexity itself, since a variety of packages are required to construct the testbench, to form a useful WAVES definition for use with a VHDL design.

The packages indicated as the *User-Developed* packages in the figure form the WAVES dataset. Within this dataset are two categories of packages: *general* ones and *test-specific* ones. The general ones serve to define waveform shapes (**Frame_Sets**), waveform logic levels (**Logic_Value** and **Value_Dictionary**), and a "shorthand" coding notation (**Pin_Codes**). We place these in the "general" category because we may re-use them over and over again for our test simulations. Conversely, the test-specific ones are peculiar to a particular test, where we need to specify particular device pins (**Test_Pins**), identify our files through particular header information (header file), actually generate the needed waveforms (waveform generator), and provide pattern or truth-table data that defines the test (external file).

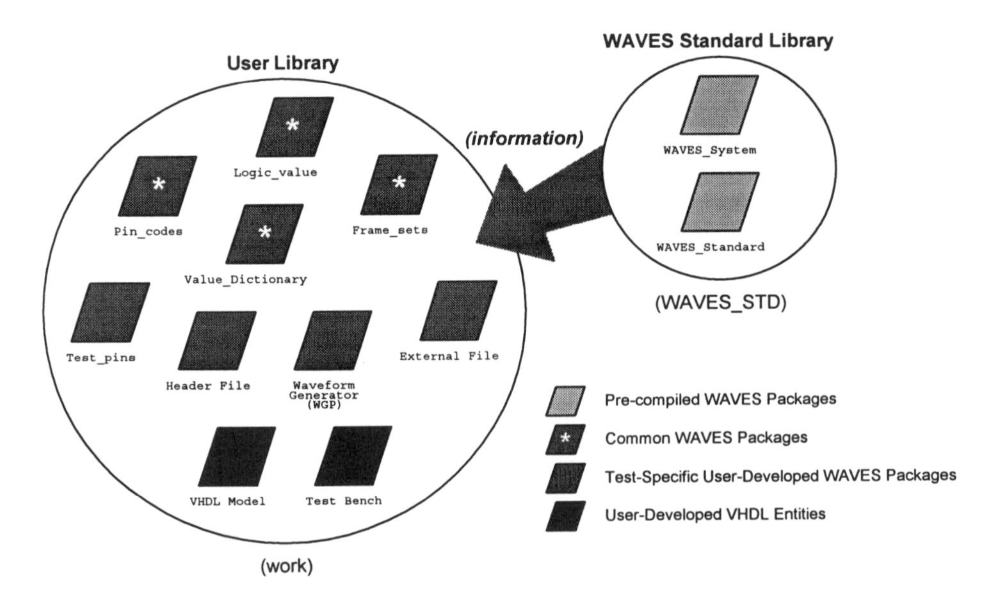

Figure 2-2. WAVES Library and Package Complexity

We construct these WAVES dataset packages and capture their relevant information for the WAVES testbench through conventional VHDL practice: the LIBRARY and USE statements which permit us to access *libraries* and *use* the package information within them. We perform our dataset construction in a work library (User Library in the figure) and obtain the information we require in the dataset packages from the **WAVES_System** and **WAVES_Standard** packages in another library, the WAVES Standard Library. Similarly, in forming the WAVES testbench, we access the VHDL model entity and the waves dataset packages through these USE statements. Hence, the user-developed WAVES dataset, along with the VHDL model and certain other utilities, constitute the WAVES testbench to ultimately be compiled for simulation.

We immediately see the inherent complexity in the library and package management required to construct our test bench. There are many sources of disparate, WAVES-specific information necessary to accomplish this construction. Hence, the lack of some standard libraries and tool support for easing this complexity were identified as weak points in achieving widespread WAVES utilization. In order to utilize WAVES, we had to be well versed in its syntax, semantics and usage, so that we could access, create, and manage all of the necessary libraries and packages to implement the testbench. Realistically, this was a problem. The effective application of a test vector standard should not require a user to become expert in that standard, just as those who test hardware need not be expert in the designs of the test equipment they use. To resolve this problem, and to make WAVES useful to the design and testing communities, some supporting libraries were developed to mitigate the complexity for the average user, as we illustrate in Figure 2-3.

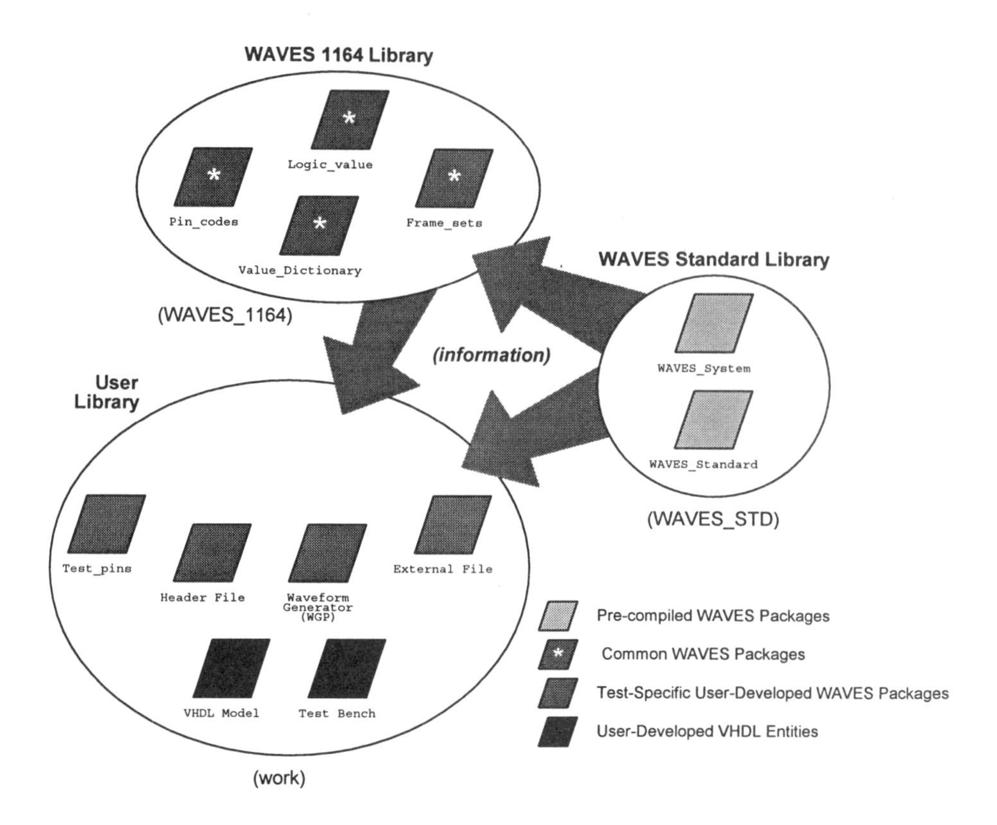

Figure 2-3. Complexity Reduction for WAVES

The figure illustrates where library elements were developed to reduce the learning curve and to provide a set of useful functions, called WAVES Common Packages, that will meet much of our needs in creating the WAVES dataset and testbench. These common packages, which reside in a WAVES 1164 Library, serve to fulfill the *general* category of user-developed WAVES, thus precluding the need for us to start from "scratch" to implement the re-usable packages within the dataset. This library, which also includes some testbench creation utilities not indicated in the figure, provides a pre-defined environment for common applications of WAVES and VHDL, and we intend that these packages will eventually be integrated into the WAVES Language Reference Manual (LRM).

In ***summary of our WAVES and VHDL integration overview***, we note that the fundamental WAVES capability benefits from certain library support to provide *more effective WAVES application.* Throughout this text, we will describe these libraries and provide both instruction and examples for their application, in detail. As for the library of WAVES Common Packages, it is the essence of the support and warrants some preliminary explanation, which we provide next.

2.3 WAVES Common Packages (the WAVES 1164 Library)

In order to appreciate the significance of the WAVES 1164 Packages we just outlined, it is useful to consider the requirements imposed upon the WAVES testbench. In using WAVES, we basically wish to create the WAVES dataset, and integrate it with the VHDL model to be tested, so that we may compile and execute that combination under simulation to obtain the performance information necessary to evaluate our design. Our points of connection to the VHDL model are called ***ports***. Hence, we need to configure the WAVES testbench with signals to connect it to the VHDL model through its ports. These port list signals are ***waveforms***, which consist of a pattern of ***logic values*** expressed as a shape which is defined as a function of time-based edge transitions. Hence, our requirements are to assemble logic values into waveforms and configure testbenches to port these waveform signals to and from the VHDL model.

To facilitate creating the WAVES dataset and testbench, we have three different sets of library functions, all of which have been developed to support the WAVES-VHDL testbench configuration. These library function sets precisely address the requirements we just defined, and consist of: (1) ***IEEE STD 1164-1993 logic values***, (2) ***waveform shapes***, (3) ***pin codes***, and (4) ***testbench utilities***. We provide a preview of these sets here to complete our background overview, and elaborate upon them more specifically in the later chapters.

2.3.1 WAVES 1164 Multi-Valued Logic (MVL) System

This library set addresses the requirement to define logic values, and was created to support a WAVES standard logic value system. It consists of the **Logic_Values** and **Value_Dictionary** packages shown previously in Figures 2-2 and 2-3. These WAVES packages define a logic value system based on the IEEE STD 1164-1993 Multi-Valued Logic (MVL) System. They provide a library of reusable elements for WAVES users who are developing VHDL models that are compliant with the IEEE standard logic package. WAVES logic values define the events (logic level transitions) that occur on the waveform signals generated by the WAVES dataset. This logic value system is almost identical to the Standard 1164 logic values. However, they also account for the waveform signal direction, which is required to generate a self-monitoring testbench.

2.3.2 Waveform Shapes

A second set of library functions addresses the waveform requirement, and was created to establish a set of basic waveform shapes (formats), that may be used to construct complete, complex waveform descriptions for all of the input and output signals for a given model or ***Unit Under Test (UUT)***. As we shall detail beginning in Chapter 3, a WAVES waveform is comprised of three basic elements: (1) patterns (the truth table data), (2) frame formats (the waveform shape), and (3) timing (the edge transition or ***event*** points in time). These three elements combine to create the waveforms for data input to the model (the drive data) and the data expected on the output of the model (the expect or compare data). A set of functions that generate common shapes, the **Frame_Sets** package of Figures 2-2 and 2-3, was constructed to simplify the waveform generation process and provide a link to ATE (Automatic Test Equipment) utilization. The library functions are used to specify the format and timing values for the waveform.

2.3.3 Pin Codes

The third set of library functions, indicated by the **Pin_Codes** package in Figures 2-2 and 2-3, is basically a time-saving indexing notation. It contains a set of particular single-character designators, called pin codes, which provide a shorthand notation for referring to frames of time-sequenced logic levels, which are segments of the waveform. Again, we will see how these are utilized in detail in later chapters.

2.3.4 Testbench Utilities

The fourth set of library functions supports the testbench configuration requirement, and was developed to provide the integration support for using WAVES in the VHDL simulation environment. This library of functions, which we do not show explicitly in Figures 2-2 and 2-3 as it applies more to testbench creation than to package management, provides the means to connect the WAVES port list signals to the model and evaluate the model response for compatibility to the WAVES expected response. This library can reduce extensive hand development of the testbench. The functions within have been developed to support any testbench with uni-directional or bi-directional pins.

In ***summary of our Waves Common Packages***, we see that they indeed satisfy the requirements to enhance and support creating a WAVES dataset for inclusion within the WAVES testbench. As this is simply a preview of these concepts, we will elaborate further on their use in later chapters.

In ***summary of Chapter 2***, we have presented the history and background of WAVES in sufficient detail to provide valuable insights for the remaining material. We have outlined the conception and motivation for WAVES, overviewed the fundamental WAVES and VHDL integration concept, and previewed the tool and library support we can obtain. In Chapter 3, we outline some technical foundations regarding waveforms, upon which we will continue to build in subsequent chapters to provide a complete understanding of WAVES, its applications, and its benefits.

CHAPTER 3. WAVEFORM CONCEPTS

An overview of our waveform terminology and conventions

Having gained an appreciation of the history and background of WAVES, we are now ready to begin outlining its application in earnest. In this chapter, we lay the foundation through an overview of the ***waveform terminology and conventions*** we will use throughout this text. Although we touched briefly upon some of these in Chapter 2, here we provide the complete familiarity necessary to move fluidly through the explanations and examples we present later. Our foundation consists of but *a few, simple waveform concepts* which are really *all we need to grasp,* to fully understand the utility and application of WAVES.

This chapter also begins what we are calling the *basic tutorial* portion of our text, which includes Chapters 3 through 7. In this portion, we will describe basic waveform concepts and demonstrate how to develop waveforms using a simple building-block approach, capture them within WAVES elements, combine these elements into a complete WAVES dataset and WAVES testbench, handle site-specific aspects, and attach and execute the combined WAVES and VHDL descriptions in support of our design or testing problems. Through this basic tutorial portion, we concentrate upon the construction and implementation aspects, with fairly simplistic timing relationships. Then, following the basic tutorial portion, we introduce the realism of more complex timing as well as additional utilization issues, in the subsequent chapters.

In this chapter we cover three principal topics. First, we provide an overview of the WAVES concept of a waveform, and particularly how it works as a design and testing specification. Next, we describe the characteristics and structure of the waveform, and the building blocks we use to define a sequence of time-ordered events specified across a set of signals. Finally, we briefly preview how to compose an entire waveform using the building block concepts. Throughout the chapter, we work in the conceptual domain, providing the general foundations of WAVES waveform creation, unencumbered by the nuances of the specific WAVES and VHDL language constructs which we describe in Chapter 4 and beyond. As we mentioned earlier, we only need to understand a few simple concepts to completely understand WAVES, and the VHDL specifics will easily fall into place later. We now begin with our waveform concepts.

3.1 Waveforms

In the WAVES context, our waveform represents *stimulus and response* information for a Unit Under Test (UUT - a physical device or VHDL model). Typically, the intent of a waveform is that when the UUT is driven with the stimulus information in the waveform, its response matches the response information contained in the waveform. More generally, the intent of a waveform is that the stimulus and response information within the waveform cooperate with the stimulus and response information from other sources. Then, when the UUT is driven with the stimulus information from all sources, its response matches the joint waveform response. In this sense, the waveform can be viewed as a *design and testing specification.* For example, Figure 3-1 shows a typical, general test environment, not necessarily a WAVES-specific one, which utilizes these waveform concepts.

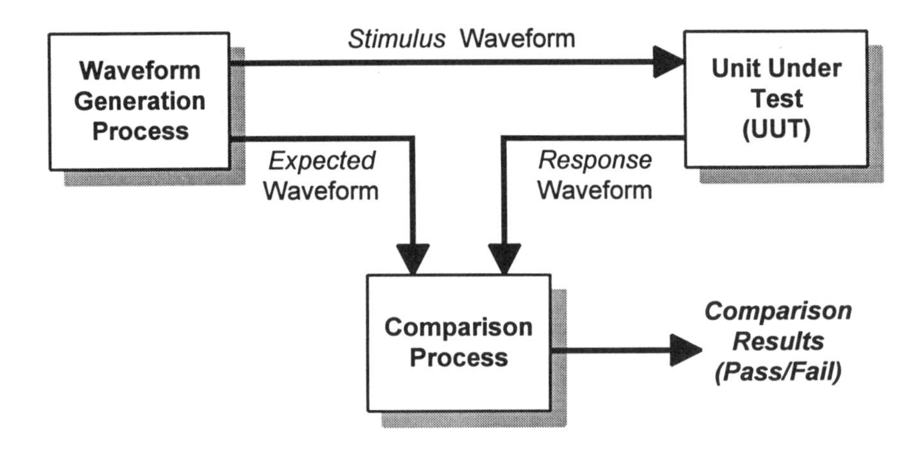

Figure 3-1. A typical test environment illustrating the waveform concept

We should note, however, that a particular waveform does not have to include *all* stimulus and response information needed by a UUT. For example, a waveform may contain only stimulus or only response, or only the UUT response and waveform expected response needed to compare during certain "critical" periods. Hence, we require some amount of flexibility in defining the contents of our waveforms - which WAVES provides us, as we shall elaborate in later chapters. **The point here is simply that a waveform contains some portion of the total stimulus and response information for a given UUT.** Next, we define that information.

3.2 Characteristics and Structure of the Waveform

In this section, we describe the composition of a waveform in the context of WAVES, but without the language or structural aspects of WAVES and VHDL. Our discussions begin in a top-down fashion, which ultimately leads us to some WAVES building-blocks and how we manipulate them. We first define how a ***waveform*** may be decomposed into fragments of time called ***slices***, and then how the waveform or its slices are comprised of individual signals. Next, we describe how the portions of the signals within these time slices, which we call ***frames***, serve as the fundamental, atomic units of WAVES: the waveform building blocks. Finally, we overview how we may organize and manipulate these building-block frames, within certain WAVES constructs called ***frame sets and frame set arrays***, to effectively use the flexibility and versatility provided by this waveform building-block approach. Once we understand slices, frames, frame sets, and frame set arrays, we have a complete conceptual understanding of WAVES. That's really all there is to it.

As we mentioned in the introduction to this chapter, we are not going to get into WAVES and VHDL language issues here. Although we describe certain WAVES terms and constructs, our discussions here focus upon a conceptual view of how waveforms may be organized and handled using the building-blocks. Here, we describe the rationale for this building-block approach as well as the capability it provides. In later chapters we will describe the semantics, syntax, file structure, and other details regarding how the waveforms may be specifically constructed and manipulated using WAVES and VHDL. We now begin with the waveform.

3.2.1 The Waveform

In WAVES, we consider a ***waveform*** to simply be a collection of time-dependent logic-level transitions, or *events,* which have some significance in the context of the test pins of some UUT, as we illustrate in Figure 3-2. Here, each series of events at a given test pin represents the time-ordered sequence of logic level changes which occur over some time duration - a signal. These may include signals we wish to generate to stimulate the UUT, as well as signals we expect the UUT to produce, which we will compare to the actual signals the UUT produces to evaluate its functionality.

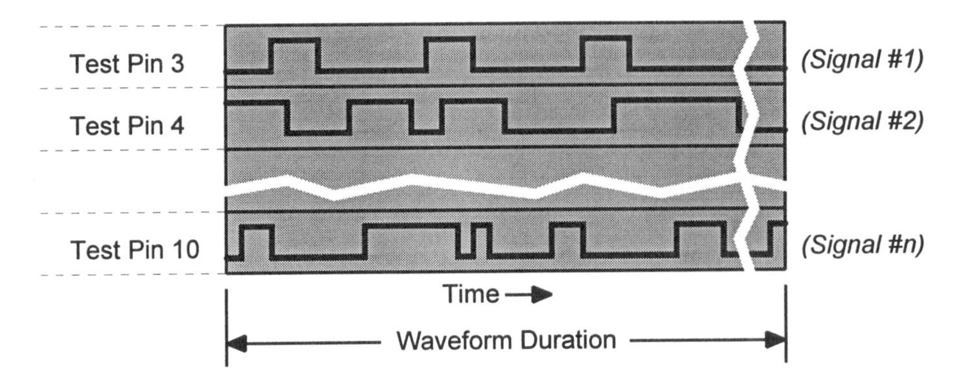

Figure 3-2. The Waveform - a set of signals over some time duration

To summarize our waveform definition, the waveform is simply considered to be a set of signals upon test pins which exists over a specified time duration. Obviously, we need to specify these signals explicitly in both time and logic level value and associate them with particular pins of the UUT, in order to stimulate the UUT or compare them with UUT-produced signals. To move in this direction, we now begin our waveform decomposition process, by taking a closer look at the slice concept.

3.2.2 Slices

The notion of ***slices*** is exactly as the name implies - slicing the waveform and its integral signals into segments of time. Although WAVES provides the flexibility to slice our waveform using any and as many time intervals we choose, it also imposes a constraint to ensure intra-waveform signal timing consistency: *the slice intervals apply uniformly across all signals in the waveform.* For example, we might decide to slice the waveform previously illustrated in Figure 3-2 using equal-time intervals which happen to be the repetition period of signal #1. This would result in the slicing depicted in Figure 3-3.

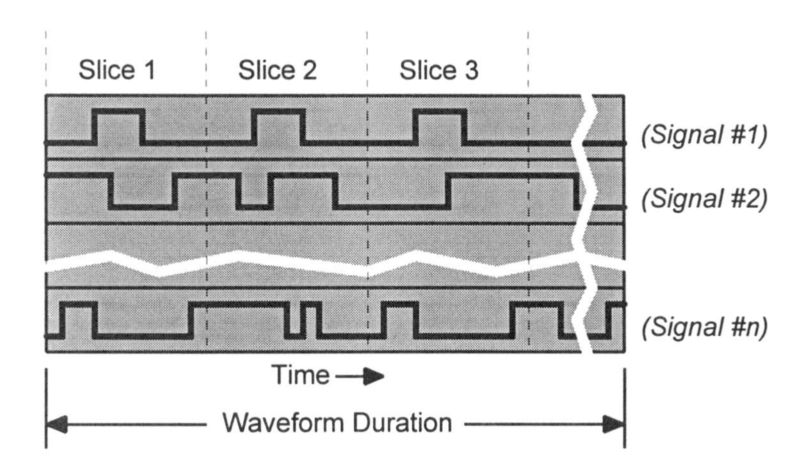

Figure 3-3. Slicing the waveform

As the figure shows, such a slicing would permit us to use repeated sequences of level-transition events throughout the duration of signal #1, but would not afford us the advantages of such repeatability in the other signals. Hence, we see that the selection of the slice interval should be done thoughtfully, and according to the particulars of the waveform we are working with. At this level of decomposition, we begin to see the ***building-blocks*** emerge as the *signal segments defined by the slice boundaries*, and we have the notion that a prudent selection of these boundaries will permit us to take advantage of repetitious event sequences.

In ***summary of our slice concept***, we see that it provides the means to time-segment our waveform and its signals. By constraining us to apply the slice intervals consistently across all signals in the waveform, it also provides us with a consistent view of the waveform.

3.2.3 Signals and Events

Signals in WAVES are exactly the same as signals in VHDL, or any other digital context. WAVES signals, however, are used to convey sequences of logic levels for stimulating the UUT and for asserting the response that we expect from it. Signals, for our discussion, can simply be thought of as sequences of "events." In this section we present and discuss the WAVES concept of the event, and compare and contrast WAVES events with the VHDL concept of an event. Here, we venture a little further into the realm of VHDL than in other sections of this chapter, because we wish to make the differences in "events" quite clear. As we will see, the sequencing, or scheduling, of these events adheres to the semantics of VHDL signal assignment.

A WAVES event is an *event time-logic value pair*, as we illustrate in Figure 3-4. The event time component may be in whatever units are appropriate to the particular design or test we are conducting. The event time is used to describe the placement, or scheduling, of the "event" on a signal with respect to the beginning of the current slice. The logic value (logic level) component may be chosen to model the accuracy or constraints of a particular simulation or test environment. Logic values, as we would expect in digital systems, describe the discrete value (such as '0' or '1') of the event. WAVES logic values do not describe physical voltage levels, as such detail is beyond the scope of WAVES.

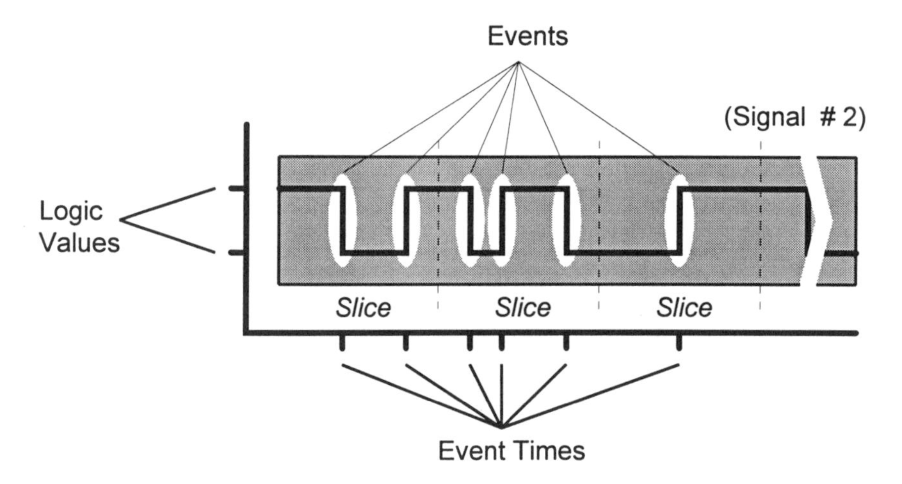

Figure 3-4. The signal - a sequence of events

A waveform, then, is constructed by scheduling events on the signals of the waveform on a "per-slice" basis. As we described in Section 3.2.2, a slice is a cross-section of the entire waveform that represents time ordered sequences of logic values for each signal. WAVES refers to all of these (time, logic level) pairs as "events" without regard to their effect on the waveform. By contrast, VHDL distinguishes between assignments to a signal that change the value of the signal and those that re-assert the same value. In VHDL, the former is called an event, while the latter is called a transaction. In VHDL parlance a WAVES event is called a transaction, since the scheduling of a WAVES event on a signal may or may not cause a VHDL event (value change). All WAVES events are applied to the signals of the waveform using the VHDL transport delay semantics. As a consequence of this, not all WAVES events will generate VHDL events. Why and when this is the case will become clear later in this chapter.

In summary of our signal and event discussion, we now see that signals are sequences of events. Events are composed of two components, an event time and a logic value. We also see that event scheduling adheres to the VHDL semantics of signal assignment. It would be difficult and cumbersome to enumerate all the event times and all the level changes explicitly for every signal in the waveform. Digital signals, however, are inherently repetitive - exhibiting certain sequences of level changes over and over. We would like to take advantage of this repetitive nature when describing the placement of events on the entire waveform. In the previous section we decomposed the waveform into slices of time. We will see in the next section that WAVES provides a mechanism for grouping events to form common, reusable "frames." We will also see that these frames may be applied to the signals of the waveform repetitively, thus eliminating the difficult and cumbersome task of explicitly enumerating every event on every signal.

3.2.4 Frames

Frames are the *fundamental building blocks* of WAVES and its *core concept*, and WAVES provides us all the flexibility we need to effectively manage them. Frames are easy to understand, create and manipulate, and are the essence of what makes WAVES a particularly helpful and useful tool. In this section we provide the details of this concept by concentrating upon the frame as an individual entity, briefly postponing the issues of how we manage and manipulate the frames to our discussions of *frame sets* in Section 3.2.5.

To begin, we already have a pretty good idea of what a frame is, from our discussions of slices, events, and logic levels in the previous section. But let's summarize these notions in the context of a single frame definition. A ***frame*** is defined as a segment of a signal between slice boundaries, which contains the events (time and logic level pairs) of the signal in the slice. In other words, a frame defines a particular, time-ordered sequence of logic-level transition events. The frame exhibits temporal flexibility, in its own duration, through the manner in which we slice the waveform. For example, if we divide the waveform of Figure 3-2 into a few, multiple-event frames we obtain the frames illustrated in Figure 3-5, for Signals #1 and #2.

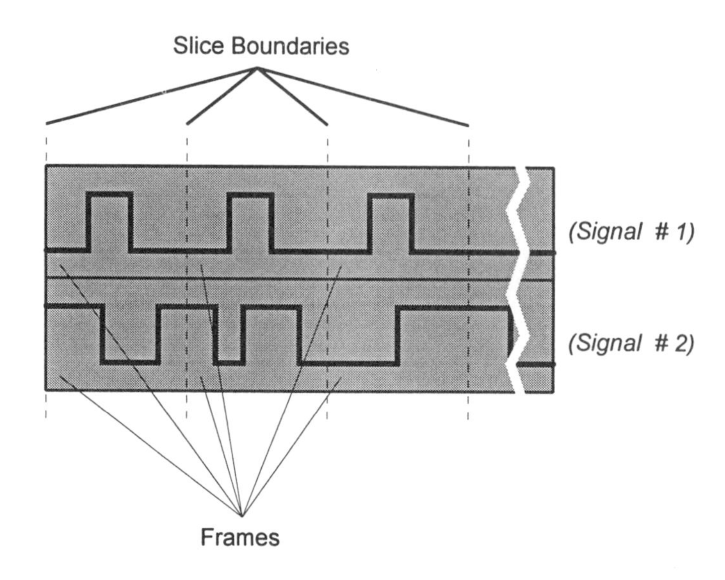

Figure 3-5. Slicing to obtain a few, multiple-event frames.

Conversely, if we slice in a finer-grained and non-uniform manner, we can obtain many frames containing single events or even no events, as Figure 3-6 illustrates, using the same waveform signals previously illustrated in Figure 3-5.

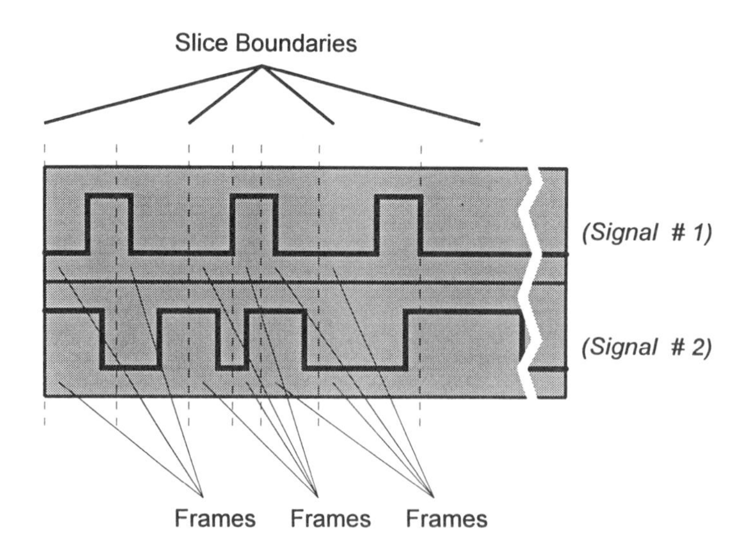

Figure 3-6. Slicing to obtain many, single-event (or no-event) frames

The point here is that of flexibility - tempered by intelligent selection. We can define our slices as we choose, but the fact that they must apply consistently across all signals of a waveform may precipitate no-event frames in some signals, due to the choice we made to capture certain events in others. We will discuss more of the nuances of effective slice interval selection across multi-signal waveforms, and provide some examples in later chapters. For now, we need only understand that this flexibility exists, and that it is quite simple to define and accomplish the slicing.

The flexibility of WAVES pervades into the frame itself. Although we consider the frame to be the basic "atomic unit" of WAVES, we can also manipulate the "subatomic particles." That is, we may specify the intra-frame dimensions of the signal, as event times and logic values of the various levels, in any manner we choose. The logic values are absolute, typically in accordance with a standard logic value definition system we shall describe in later chapters. Here we may choose a resistive high or low, or a forced high or low, as well as several other level values. The event times are referenced to the beginning of the frame, which is defined by the slice boundary at that point. Hence, the slice event timing relies upon the time synchronization provided by the uniform, inter-signal slice interval definitions we discussed earlier, to define actual times for the events. Figure 3-7 illustrates these intra-frame flexibility features.

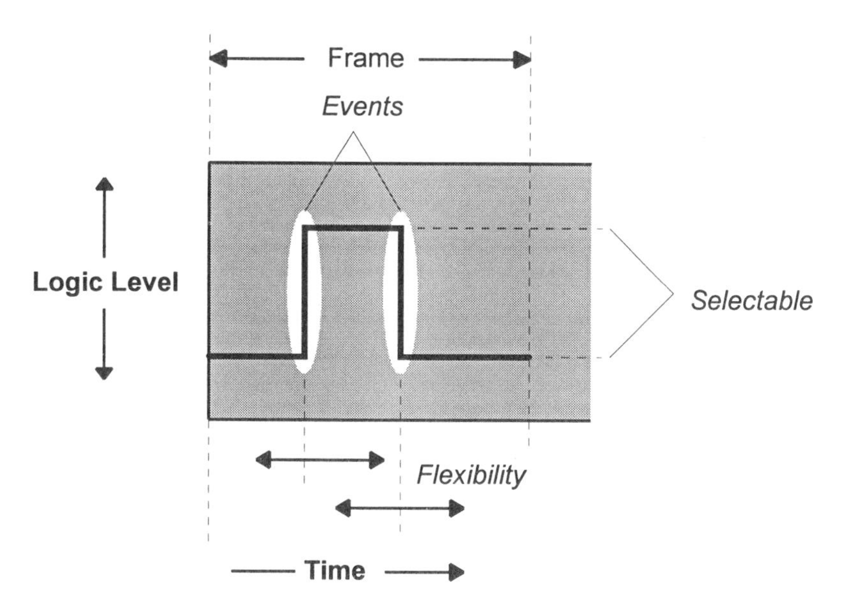

Figure 3-7. Signal dimension flexibility within the frame

At this point, we see that the flexibility of WAVES provides the means to manipulate waveform dimensions within frames, permitting us to develop signals and waveforms basically any way we choose from the frame building-blocks. We simply construct the frames the way we want them, assemble them into multiple-signal slices and sequentially construct waveforms composed of time-ordered, multiple slices. Now, the final frame-related topic we need to cover is a that of some definitions - we need to be explicit about what constitutes *different* frames.

Basically, the notion of different frames is a easy one, which we will illustrate through some examples. *Different frames* simply have one or more *different dimensional values* - different logic levels or event times - and *multiple instances of the same frame do not have these dimensional differences.* For example, if event times and logic levels are *precisely* the same among several frames, as we depict in Figure 3-8, these are considered multiple instances of the same frame.

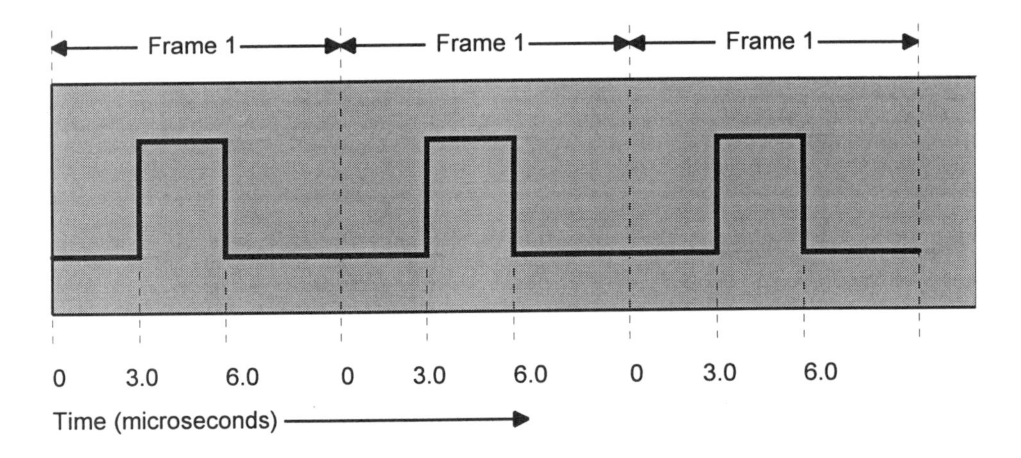

Figure 3-8. Multiple instances of the same frame

It follows, then, that if event times or levels differ among frames, even though they contain the same sequence of events (such as logic level transitions 0-1, then 1-0), they are considered separate and different frames. We illustrate this concept in Figure 3-9, where differences in logic levels or event times among the three frames render them as being different. Here, and in subsequent graphical renditions of waveforms, we use the *solid line to denote a hard, or forced logic level*, and a *dashed line to denote a soft, or resistive logic level.* In the figure, frame 3 differs from frame 1 because it has a soft high during the active portion of the pulse.

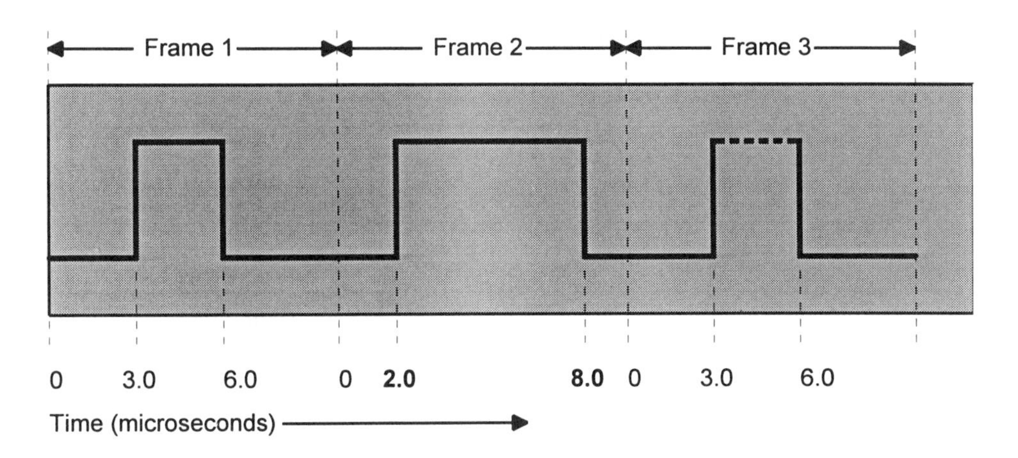

Figure 3-9. Three different frames, due to dimensional differences

An important aspect here is that within the structure of WAVES, while the frames include actual logic level values, their event times appear simply as designators. For example, a frame which exhibits a pulse which starts with a logical low value at 0μs, goes high at the first event time (t1 = 3μs), and then returns to low at the second event time (t2 = 6μs), simply includes these event designators (t1, and t2). The actual time values (3μs and 6μs) are associated with the frames later, in the parent *frame set array*, which we describe later in this section.

In ***summary of our frame definition and discussions***, we now understand the key features of these basic WAVES building-blocks. Frames have the waveform dimensional flexibility we need to specify the particular signal behavior we require. Because they are bounded by waveform slice boundaries which provide the time references, the frames are temporally consistent across all waveform signals. They are considered to be different frames if they contain different dimensional properties, and multiple instances of the same frame if they do not. Now we are ready to look at how we can organize these frames and associate them with the UUT, through the use of *frame sets* and *frame set arrays*.

3.2.5 Pin Codes, Frame Sets, and Frame Set Arrays

A ***frame set***, as its name implies, is a *collection of frames for a particular purpose - to create a particular signal or group of signals*. We may define it as the set of all frames from which we will compose a signal to (1) apply to a particular pin of a UUT, or (2) compare with a signal coming from a particular pin of a UUT.

We illustrate this through an example. Assume we require two signals, let's call them Signal #1 and Signal #2 (not the same Signal #1 and #2 we used in the previous figures of this chapter), to stimulate or compare with signals from some UUT. Also assume that these signals are to be constructed from certain frames of logic level events, as we illustrate in Figure 3-10. In this example case, Signal #1 is a sequential repetition of four different frames, and Signal #2 is a sequential repetition of three different frames. Hence, we require a frame set with four entries for Signal 1, and one with three entries for Signal 2.

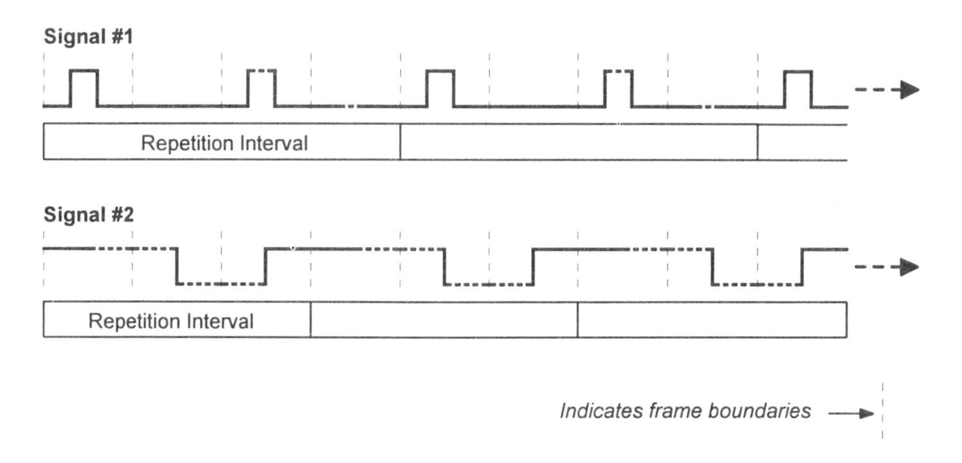

Figure 3-10. Two example waveform signals composed of repetitious frames

Let's assume we wish to construct these frame sets, as Frame Set #1 for Signal #1 and Frame Set #2 for Signal #2. First of all, we need a mechanism to select frames from the frame set to apply to the signals. In WAVES, this mechanism is the ***pin code***. Let's begin with Signal 1. For each sequential frame, it appears to be a pulse shape, basically starting at a hard (forced) logical low, and exibiting one of four characteristics in its active portion: (1) a hard (forced) high, (2) a hard (forced) low, (3) a soft (resistive) high, and (4) a soft (resistive) low. We will choose pin codes which let us describe the nature of the logic level within the active portion of the pulse, which also happen to be consistent with the IEEE 1164-1993 Standard Logic Values system. Here, we can designate our hard high as a *1*, our hard low as a *0*, our soft high as an *H* and out soft low as an *L*. Now, although we only need three pin codes to describe the three-frame repetitious characteristic of Signal #2, we are required to use the same (entire) pin code set for all frame sets. In this signal, we have a mid-slice transition among three levels: (1) a hard high, (2) a soft high, and (3) a soft low, and the level prior to the transition event is whatever it was in the previous slice. These three levels also appeared in Signal #1, but the shape was a pulse instead of a mid-slice transition. Hence, we simply use three of the four pin codes we chose for Signal 1 to apply to the three frames we need, and ignore the fourth pin code, letting it be a blank or empty frame. The resulting frame sets and their pin codes, for this example, are shown graphically in Figure 3-11.

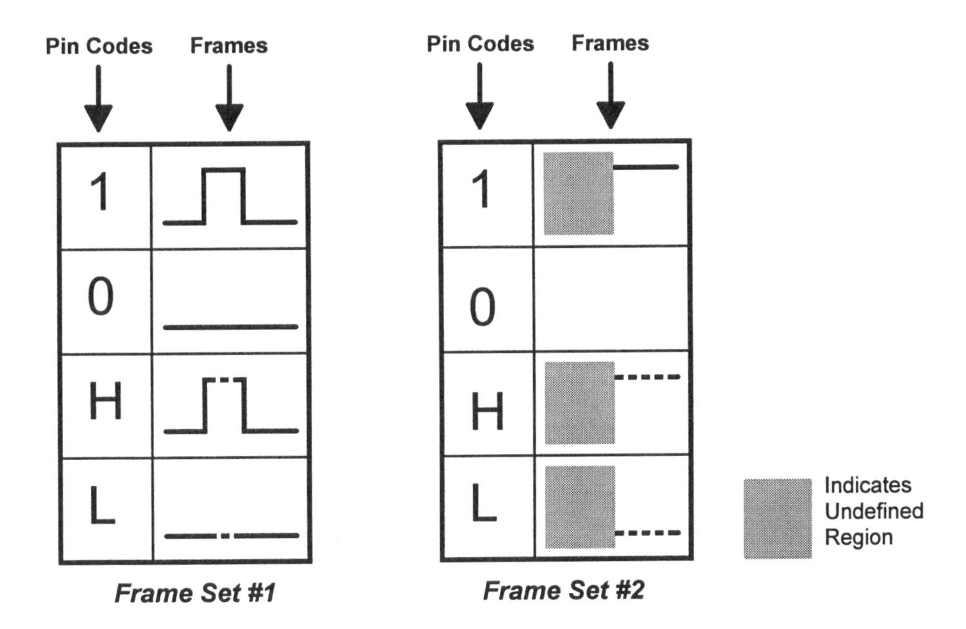

Figure 3-11. Frame sets for the two example waveform signals

To briefly summarize the shorthand notation called pin codes, they are simple, *single-character identifiers* which designate a particular frame within the set. In choosing the pin codes, we again have complete flexibility, and can basically use any single-character designators we choose. However, a restriction applies here as well - that we must use the same set of pin codes to define all frame sets for a particular test.

Here, we see the convenience and flexibility we have within the constraint of using the same pin codes for each frame set. Our Signal #1 may be defined as the Frame Set #1 pin code sequence: 1, 0, H, L; repeated through the duration of the waveform. Similarly, our Signal #2 may be defined as the Frame Set #2 pin code sequence: H, L, 1; also repeated through the waveform duration. Since we never index the frame corresponding to pin code 0 in Frame Set #2, we do not care what frame it defines, or whether or not it defines any frame at all. In both frame sets, the pin codes actually serve two purposes: to (1) specify a logical level and to (2) define the nature, or *format* of the signal in the frame. There are many useful formats we may use, as we will explain in Chapter 4. Here, we used a "pulse" format for Signal 1 and what we call a "non-return" format for Signal 2, implying a transition at some inter-slice event time and holding that level until the same event time in the next slice. The ease and simplicity of this concept is apparent. The pin codes define the levels and the shapes indicate where the levels apply. Obviously, with a few levels and shapes we can define myriad waveform segments, and that is precisely the power of WAVES - a few building blocks to produce anything we need.

Such flexibility and versatility naturally leads to some questions surrounding effective utilization. Given that we *can* designate and organize our frames in any manner we choose, how *should* we do it? For instance, if we decided to slice the waveform of Signal #1 in Figure 3-10 differently than before and used only pin code 1, as we illustrate in Figure 3-12, we could obtain a frame set with the same basic shape in all the frames. However, since their event times differ, these would still be different frames, as the figure illustrates. Are we actually more efficient by choosing frames of similar shapes? Alternatively, we could slice the signal very coarsely to use a very small number of frames with many events, or very finely to use a large number of single-event frames. Which is more effective?

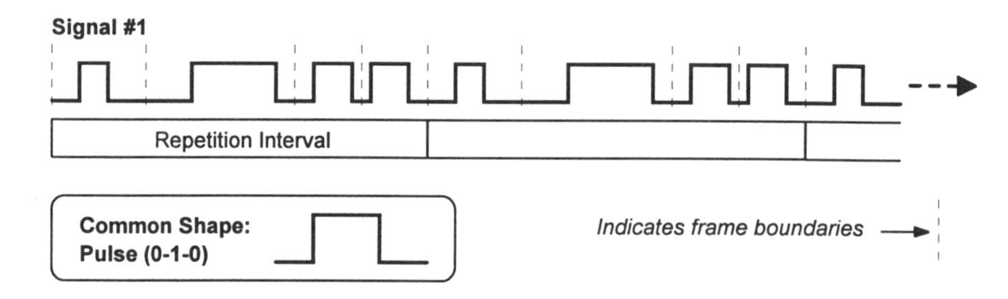

Figure 3-12. Different slicing for common-shape frames

The answers to these, and other questions surrounding effective utilization lie within the particular applications and the nuances of the WAVES and VHDL language constructs, which we will explain in later chapters. The important message here is that complete flexibility in definition and organization of frames exists, and our only constraints are that we must slice all signals within the waveform identically, and we must use the same set of pin codes for all frame sets in a waveform for a particular test. Now let's move on to the frame set array.

The ***frame set array*** serves three purposes. First, it captures all the frame sets we require for the signals within a particular waveform. Second, it associates the frame sets with particular pins of the UUT. Finally, it applies the specific time values to the event time designators within the frames. For example, assume we wish to apply Signal # 1 to two pins of some UUT, say pins 6 and 9. Then we wish to compare the signal we receive from UUT pin 8 to Signal # 2. Obviously, there is an association between pins and signals we need to make, and the frame set array, depicted graphically in Figure 3-13 serves that purpose. Within the frame sets, as we described earlier, the pin codes identify certain frames for the particular pin, and they index logic values for the events within the frames. Now, within the frame set array, the pin codes also index event times. Hence, for a particular pin, if pin code "1" designates a frame with two logic level events, it also indexes two time values, and so forth.

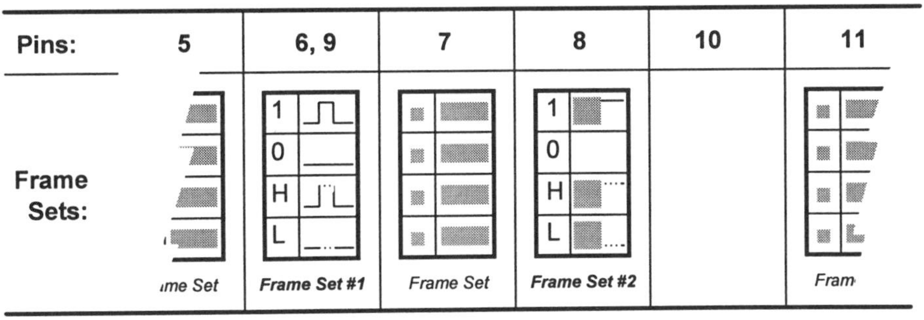

Figure 3-13. Pin and frame set associations in a frame set array

Within the frame set array, the pins are the keys. Each particular combination of frame set and event times for a UUT pin represent a unique signal. If that unique signal is required for multiple pins, the frame set and event time values may be simply be repeated for that pin. Where we require similar shapes with different times, we can easily re-use frames or frame sets, and provide different event times in the frame set array. This organization, which separates time and events, is of significant value to effective WAVES utilization, as we shall detail in later chapters.

In ***summary of the frame set array***, we see that it is a collection of frame sets for the purpose of a particular test, which associates each UUT pin of interest and event time values with a particular pin-code-indexed frame set. At the frame set array level, we have complete temporal and logic value definitions of all frames for all pins of the UUT. Furthermore, if the same set of frames is to be used with several pins of the UUT, the same frame set may be designated for multiple pins (as for pins 6 and 9 in the figure). The frame set array then includes the *frames sets for all signals in the waveform we require for a particular test*, which is a *subset* of all possible frame sets.

In summary of this section concerning the ***Characteristics and Structure of the Waveform***, we have looked at how the waveform is considered to consist of time-sliced signals which, in turn, are composed of frames of logic-level transition events. We have also illustrated how these frames may be manipulated, using frame sets and frame set arrays, such that a shorthand pin code notation can define the frames of logic level transition events and their times, to be applied to pins of a UUT. At this juncture, the missing element is sequence, since we have established which frames apply to which pins of the UUT, but have not defined the order in which they must appear or their actual slice times. We need a mechanism to index the frames of logic level transition events over a particular sequence of time slices, in order to produce the time-based signal and waveform. The mechanism for accomplishing this sequential ordering task is called an *external file*, which we describe in the context of composing the waveform, in the next section.

3.3 Composing the Time-based Waveform

In this section, we preview how the frames, frame sets, frame set array, and external file we discussed in the previous section work together to create the waveform and its signals, with both logic level and actual time dimensions in place. Remaining consistent with the focus of this chapter, our descriptions here are conceptual, and not set in the specifics of WAVES or VHDL language and structural constructs. We provide such details in later chapters. Also, we provide only a cursory view of the external file here, restricting ourselves solely to how it functions to provide the sequenced timing information to the waveform and signal composition process. The complete depth and breadth of the external file constitute sufficient volume for a chapter of their own, to which we attend in Chapter 7. We now begin with the sequencing contributions of this external file.

3.3.1 The External File - A Simple Array of Duration Data

For our purpose here, we need only concern ourselves with the aspects of the ***external file*** which permit us to apply frame order and slice timing information to the waveform signals. We can think of the external file simply as *an accumulation of sequences of pin codes for each UUT pin, indexed by an ordered series of time slices across the duration of the waveform.* Each particular slice duration designates a pin code for each UUT pin number. We illustrate this view of the external file graphically, in Figure 3-14, as it applies to three of the UUT pins (6, 8, and 9) for which we developed frame sets in the example of the previous section.

Cumulative Waveform Time at End of Slice	Pin: / Slice Duration Time (ns)	6 Pin Code	7 Pin Code	8 Pin Code	9 Pin Code
200	200	1		H	1
400	200	0		L	0
600	200	H		1	H
800	200	L		H	L
1000	200	1		L	1
1200	200	0		1	0

External File Contents →

Figure 3-14. Graphical rendition of a portion of an external file

Through this example we can see how the external file orders and time-tags the frames and subsequently provides the reference timing for their integral events. Recall that the sequence of frames for these two pins were designated by repetitions of pin code sequences 1, 0, H, L (for Signal #1 on pins 6 and 9) and H, L, 1 (for Signal #2 on pin 8). Here, the external file includes that repetitious sequence and also assigns a slice duration to each pin code for each UUT pin, thus designating the starting time for each frame of events as the beginning of a particular duration. Through this simple ordering of time durations, the cumulative waveform time is simply built up sequentially as each slice's duration is added to its predecessor. Again we see the buliding-block approach, this time as building up elapsed time with "blocks" of durations. Recall that the frame set array includes the event times (with respect to the slice starting time) for all events in the frame. Hence, through a combination of the external file and the frame set array, we obtain the absolute times for these events simply because the event times refer to the slice starting time, and the slices are being accumulated sequentially. Next, we present an overview of how this external file, and the other constructs we have described, work in concert to define the waveform and its signals.

3.3.2 The Waveform Composition Process

From the individual descriptions of the waveform components and the external file, given in the previous sections of this chapter, we can begin to see how the composition process works. To complete our picture, in this section we describe the composition process in terms of how the system uses these elements to create the waveform. Again, this is a conceptual view, and we are not attempting any correlation to WAVES or VHDL language constructs here. We simply present a description of how it works.

Our conceptual view of the waveform creation process is summarized in the graphical depiction of Figure 3-15. Here, we illustrate the contributions of the various elements to the completely-defined waveform. The waveform is composed of signals, associated with UUT pins, which are themselves composed of frames defined by slice duration times. Within the frames, event times define transitions among logic levels. Referring to the figure, we can see how the frame set array and external file work to provide all the information necessary to produce the waveform, block by block.

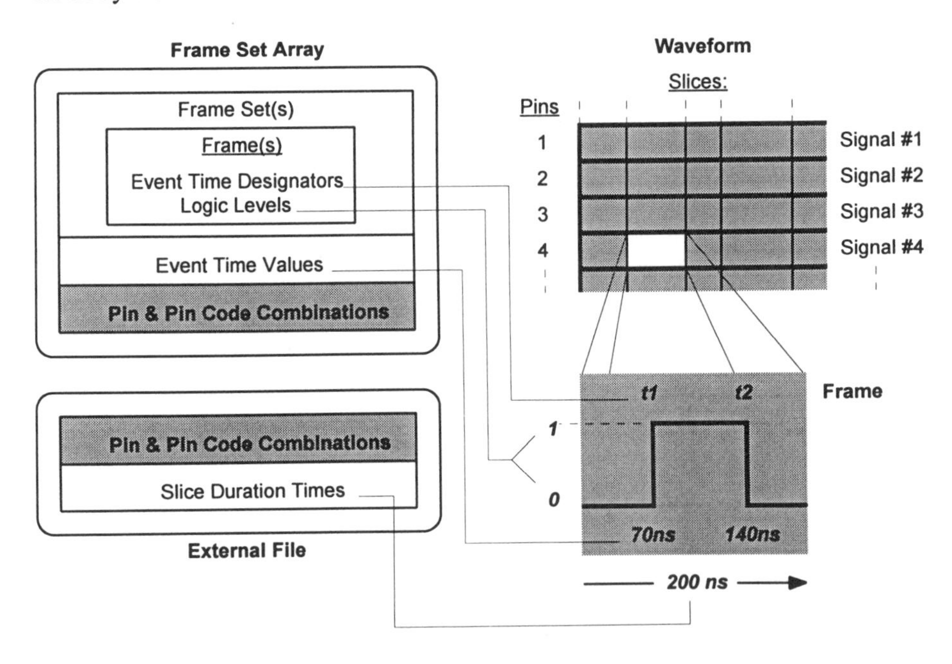

Figure 3-15. The waveform composition process

The concept is simple. Basically, the frame set array provides the logic level dimension and the frame-level (event) timing for the signals and the external file provides the slice sequencing and the waveform-level (slice) time dimension. The frame set array includes (1) frame sets of frames which contain their shape formats and event time designators, and (2) event time values, both indexed by pins and pin codes. Recall that the pin code implies the logic level as well. Therefore, any given pin and pin code combination in the frame set array defines a particular combination of shape, logic levels, and event times. In the figure, we show a particular frame required for pin 4 of some UUT in a particular slice of time. Here, the pin designation is 4, and the pin code is 1. In our standard logic convention, the pin code indicates that the logic level of the pulse will be a 1 during its active (high) portion. In the frame set for pin 4, pin code 1 must designate a "pulse" shape, indicating two event times, t1 and t2. In the frame set array, two specific times with repect to the slice starting time, 70ns and 140ns, must exist for the pin 4 - pin code 1 combination. The external file includes ordered slice duration times for the frames associated with each pin, indexed by the same pin and pin code combinations. For our case, the duration time for this particular slice must be 200ns. In summary, both the frame set array and external file provide their dimensional information in the context of the building-block frames, indexed by the pin and pin code combinations common to both entities.

Procedurally, the system simply moves through the external file according to the time-ordered sequence of slice durations. For each slice, as it is sequentially encountered, the system obtains the duration times for a signal frame, and the particular pin and pin code combination for that frame, directly from the external file. It then refers to the particular frame and event times, indexed by that pin and pin code combination, in the frame set array and frame set to obtain the logic level and slice-referenced time information for the pin associated with the frame. As the system moves from pin to pin within a slice, and then from slice to slice, it obtains the timing and logic level information necessary to completely describe the signals of the waveform, slice by slice. The waveform signals thus created and described are then applied to the pins of the UUT through a *waveform generator procedure*, or compared to signals appearing on UUT pins through a *compare function*, both of which we will describe in detail in later chapters.

To summarize the ***waveform composition process***, it uses the elements we have discussed in an ordered process of applying the building-block frames to compose the signals and consequently the waveform. By keeping the logic level and timing information somewhat separate (times in the external file and frame set array, and logic levels in the frame sets), it offers us the flexibility to manipulate these two waveform dimensions individually. Hence, we can easily use the same logic level sequences with different event timing, or the same event timing with different logic levels, which makes waveform creation, portion repetition, or composition and adjustment quite simple.

In ***summary of Chapter 3***, we first introduced the notion of how waveforms are used in a stimulus-response paradigm, to test UUTs. Next, we described the WAVES technique of waveform creation using building blocks of frames, and presented a conceptual view of how the waveform is decomposed into signals, slices, and frames to achieve the building-block paradigm. We then described, again conceptually, how the composition process works both structurally and procedurally. At this juncture, we have a good grasp of the frame-based building block approach WAVES employs to create waveforms. Now, we are ready to move out of the conceptual domain and into the realm of WAVES and VHDL, to specifically describe how this waveform composition process actually operates, through the waveform generator procedure and compare function. We begin our description by introducing the WAVES constructs which actually implement the waveform composition, in Chapter 4.

CHAPTER 4. WAVES CONCEPTS

Understanding the elements we work with

Having become familiar with the waveform terminology and concepts we described in Chapter 3, we can now begin to apply these concepts to constructing the particular waveform elements we require, using actual WAVES and VHDL constructs. Recall that in the earlier chapters, we described certain elements of WAVES as being reusable. That is, they could be applied in multiple instances, for numerous tests, to form portions of a waveform. Conversely, other elements were more test-specific, applying to a particular instance of waveform generation for a particular test. In this chapter we describe the construction process for the reusable elements of WAVES in detail, describing how to implement these foundation waveform concepts in actual WAVES and VHDL constructs.

Our subjects in this chapter, the reusable elements, include the *logic value system*, the *pin codes*, and the *frame set*. Since we might think about how to define logic values before we concern ourselves with how to code them or apply them to frames which specify shapes of transition events, we present our topics in that order. We begin with an explanation of the logic value system which is used to define and document all the possible physical values that appear on the waveform. We utilize this logic value system to define the logic levels of the waveform shapes (often called *formats*) for a slice of time in the frame sets. Next, we describe the pin codes which comprise a list of one-character identifiers (sometimes called *legal patterns*), used as a shorthand notation for indexing frames in the frame set. As we described in Chapter 3, a one-character code within the pin codes represents a frame defined in the frame sets. Finally, we present the function and implementation of the frame sets. These frame sets associate each pin code with a corresponding frame, which defines the shape of the signal within a slice as a series of time-tagged level-transition events.

Looking ahead into some of the later chapters, these WAVES building blocks will ultimately be used to construct a complete waveform, using a waveform construction process. This process requires more than the three WAVES building blocks we address in this chapter. It requires the entire WAVES dataset which, according to the WAVES LRM (IEEE STD 1029.1-1991), is the "logical collection of data required to completely specify one or more waveforms." As we described in Chapter 2, for the dataset we need some other elements, in addition to the logic value system, pin codes, and frame sets. We also need *test pins*, a *waveform generator procedure*, an *external file*, and a *header file* to make up the complete WAVES dataset. We will describe this complete WAVES dataset including its structure, elements, and functionality, in Chapter 5. For now, we will only be concerned with the three particular, reusable building blocks of logic values, pin codes, and the frame set.

Before proceeding with the technical material of this chapter, we need to address some standards and compatibility issues. First of all, we will use the **Std_Logic_1164** logic system (IEEE Standard Multivalue Logic System for VHDL Model Interoperability, IEEE STD 1164-1993) in our WAVES-VHDL simulation environment for all the examples we present. The **Std_Logic_1164** system was chosen because it is the most widely-used multivalued logic system in digital design and simulation environments. In this chapter, we develop the WAVES building blocks based on the IEEE STD 1164-1993 and capture them in a set of WAVES packages, which are reusable, facilitate generation of the WAVES dataset, and promote the easy interface of waveform information with the WAVES-VHDL simulation environment. For our examples, these reusable packages will be captured in a library called **WAVES_1164**.

If we wish to conform to this standard, we may use the examples in this chapter directly in our datasets. Hence, they may become templates for actual applications as well as textual examples, subject to a couple of constraints. First, when designs and simulation environments require a different logic system, rather than the **Std_Logic_1164**, then we must build our own set of WAVES building blocks appropriate for these environments. This is not a severe constraint, as we provide all the information necessary to do so in this chapter, and address some of the more specific aspects of this situation in Chapter 13 when we deal with advanced topics. Second, when using these examples directly, we should note that, currently, WAVES is not compliant with VHDL-93. Hence, all the examples of the WAVES datasets and their elements we present here, as well as in subsequent chapters, must be analyzed or compiled under IEEE STD 1076-1987 (VHDL-87).

Now, having previewed the chapter and settled the standards and compatibility issues, we can begin our chapter in earnest, starting with the logic value system.

4.1 The Logic Value System

The logic value system is the first WAVES building block that we must define to support a particular simulation environment. The purpose of the logic value system is to define and document all the possible physical values that appear on the waveform. We use this logic value system to unambiguously define the logic levels which may occur. Here, we describe the two components required to define the logic value system, and provide an example package that defines a logic value system supporting the IEEE standard logic simulation environment.

Our logic value system consists of a type declaration and a function. The type **Logic_Value** enumerates the different logic levels needed to construct the waveform. The type **Logic_Value** must encompass an appropriate number of logic values, such as those of the IEEE STD 1164 package, to support a given simulation environment. Along with this logic value, we use a **Value_Dictionary** function to document the meaning of each element enumerated in the **Logic_Value** type declaration, with regard to state, strength, direction, and relevance. Simply stated, this function takes a logic value as an input and returns a corresponding **Event_Value** which consists of four attributes: state, strength, direction, and relevance. This is also known as a *4-tuple* of these attributes. We note that the purpose of the **Value_Dictionary** function is simply to document the semantics of the logic values, and is not actually used during the WAVES/VHDL design simulation.

Since the **Value_Dictionary** simply returns this **Event_Value** type, we only need a cursory understanding of it to continue our logic value system discussions. As we stated above, an **Event_Value** is a combination of four separate components: ***state***, ***strength***, ***direction***, and ***relevance.*** State and strength are common simulator terms. ***State*** represents a Boolean logic level while ***strength*** is the ability to force a logic level in the face of conflicting states from other signal sources. The direction and relevance components are event value components oriented towards testing. ***Direction*** indicates whether an event represents a value to be driven by the waveform (stimulus) or represents an expected value from the UUT (response). ***Relevance*** indicates whether or not an event value is significant to the intent of the waveform. Typically, relevance is used to indicate occasions when, although the waveform predicts a UUT output (state, strength) pair, the UUT need not output that particular value in order to meet the intent of the waveform. This level of understanding of the **Logic_Value** system will suffice for our purposes here. The reader may consult Section 6.1 of IEEE STD 1029 for more information.

Having discussed the purpose and components of the logic value system generically, we now present, as an example, the **Waves_1164_Logic_Value** package, developed by the USAF Rome Laboratory for the simulation environment utilizing the IEEE standard logic package. The WAVES code for the package declaration is given in Figure 4-1, following certain, necessary context clauses (such as LIBRARY and USE clauses). The context clauses are used to make other packages residing in different libraries visible to the current package. In this example, as well as in other code examples we present in this text, the *italicized lines* are inserted for explanation or clarification purposes, and are not part of the actual code. Recall that the purpose of this package is simply to enumerate the different logic value types.

```
-- The context clauses
LIBRARY WAVES_std;
USE WAVES_std.WAVES_standard.ALL;

-- Package declaration
PACKAGE WAVES_1164_logic_value IS

-- Define the logic values
    TYPE logic_value IS
                    (
                      dont_care,
                      sense_x,
                      sense_0,
                      sense_1,
                      sense_z,
                      sense_w,
                      sense_l,
                      sense_h,
                      drive_x,
                      drive_0,
                      drive_1,
                      drive_z,
                      drive_w,
                      drive_l,
                      drive_h );

--  Function value_dictionary declaration
FUNCTION value_dictionary( value : logic_value )
        RETURN event_value;

END WAVES_1164_logic_value;  -- End of the package
```

Figure 4-1. WAVES_1164_Logic_Value Package Specification.

Our package follows a logical progression consistent with proper VHDL structure. First, the package is placed in the context of the **Waves_Standard** package, which is required to make the **Event_Value** type visible (through use of the context clauses). More extensive use of the context clauses will be presented in Chapter 5 when we discuss the relationships and visibility among all the WAVES building blocks. Here, however, we are more concerned with the package contents.

Next, we find one type declaration and a function call. First, the **Logic_Value** enumerated type is declared, using the IEEE Standard Logic notation we depict in Figure 4-2. This enumerated type declares the names of all of the legal logic values that can be used to generate events on a waveform for the IEEE standard logic package environment. Under the **Logic_Value** enumerated type, the first logic value that is listed is **Dont_Care** which corresponds to the '-' code in Standard Logic. Then, there is a set of **sense** logic values and a set of **drive** logic values for each of the following Standard Logic codes, 'X', '0', '1', 'Z', 'W', 'L', and 'H', which also belong to the set of Standard Logic codes illustrated in the figure. In our example, "sense" implies that the value is to verify a UUT port output and "drive" implies that it is to stimulate a UUT port. Also, we notice that no logic value in our package corresponds to the Standard Logic code 'U.' This is because it makes little sense to drive a model with an uninitialized value or verify that an output of model generates an uninitialized value. Finally, the **Value_Dictionary** function, which we outlined earlier, is declared. In summary, this package simply permits us to define all the logic values we are going to use, and to declare a function to map logic values to event values.

U	An uninitialized state
X	An unknown forced logical state
0	A strong, forced logical low
1	A strong, forced logical high
Z	A high impedance
W	An unknown weak logical state
L	A weak logical low
H	A weak logical high
-	Don't care

Figure 4-2. Standard Logic Values (from IEEE Std 1164-1993)

There is a certain rationale behind grouping these logic values as shown in the **Logic_Value** type declaration, which is rooted in VHDL conventions. The logic values enumerated in our example represent three separate groupings: the **Dont_Care** group, the **sense** group, and the **drive** group. This grouping reflects the different nature and use of these logic values.

The **Dont_Care** logic value appears first in the declaration for initialization reasons. According to the language definition of VHDL, after elaboration, all signals not explicitly assigned a default value have the value T'left, where T is the type mark of the signal. This causes the post-elaboration values of each signal of the WAVES dataset to be **Dont_Care**, since **Logic_Value'left** is **Dont_Care**. The post elaboration values of all implicitly-defaulted outputs on any model which is compliant with IEEE STD 1164-1993 will be 'U', since **STD_Logic'left** is 'U'. When this is the case, the WAVES dataset effectively states that it does not care that the model may have un-initialized outputs at the beginning of the simulation. This is consistent with the way that the VHDL model will be tested.

The **Dont_Care** logic value is grouped separately from the **drive** logic values, since it makes little sense to drive the model with a "don't care." Typically, when we truly don't care, the model is driven with either a '1' or a '0', not a '-'. In fact, when a resolved signal is driven with '-', the standard logic (1164) resolution function generates an 'X'. Therefore, the function of the **Dont_Care** logic value is for stating that we do not care what the actual output of the model is at a given time. This function differs from the **sense** logic values in that each **sense** logic value indicates the value that we expect on a given model (UUT) output.

The **drive** and **sense** groupings simply reflect the two different uses for waveforms. The **drive** logic values occur on the waveform whenever a signal generated by the WAVES dataset represents a stimulus to be "driven" on an input of the model. The **sense** logic values occur on the waveform whenever a signal generated by the WAVES dataset represents the expected response of the model, to be "sensed" on a model output. This dual logic value system allows WAVES to represent the direction of the signals that make up the waveform, and is necessary information for verification.

This package is reusable for simulation and design environments that utilize the IEEE standard logic 1164 system, and is included in the library **WAVES_1164**. The source code for the **Waves_1164_Logic_Value** package, including the package body, is given in Appendix A. This package is also included on a companion CD-ROM to the book, for immediate use. The name of the file on the companion CD-ROM is "**Wav_1164.Vhd**" located in the **WAV_PACK** directory. The package body consists only of the implementation of the **Value_Dictionary** function which, as mentioned above, is simply used to document the meaning of each element enumerated in the logic value type declaration with regard to state, strength, direction, and relevance. This function is not used by VHDL when simulating a WAVES dataset. The only purpose of the **Value_Dictionary** is to document the semantics of the logic values.

In ***summary of our logic value system***, we have described the two components needed to define a logic value system. First, we outlined how the **Logic_Value** type defines all the names of all of the legal logic values that can be used to generate events on a waveform. This **Logic_Value** is used to define the waveform shapes within the frame sets, to be used for waveform construction. Next, we discussed how the **Value_Dictionary** function is used to document the semantics of the logic value names. Finally, we presented a reusable logic value package for the IEEE standard logic 1164 simulation environment which is included in the library **WAVES_1164**. Next, we look at the second reusable WAVES building block - the pin codes.

4.2 Pin Codes

The Pin Codes are the second WAVES building block that must be defined for a simulation and testing environment. As we described in Chapter 3, a Pin Code can be viewed as a one-character notation of each frame defined for a waveform. In other words, a one-character code is associated with each frame in a frame set. The frame set (which we describe later, in Section 4.3 and Chapter 7) is a collection of all frames that are sufficient to describe one or more waveforms, and it associates each pin code with an appropriate frame. Hence, we see that the pin codes provide the simple representation necessary to effectively communicate the waveform information among different design and testing environments. Given this, a slice, which is a segment of time across all signals upon all test pins, can be specified simply as a list of pin codes, with one code per test pin (signal) and each code referring to a frame to apply to the corresponding pin. In turn, a complete waveform, consisting of many sequential slices of time across many test pin signals, can be represented as a set of lists of pin codes, one after the other in time-slice sequence. We illustrate these relationships conceptually in Figure 4-3.

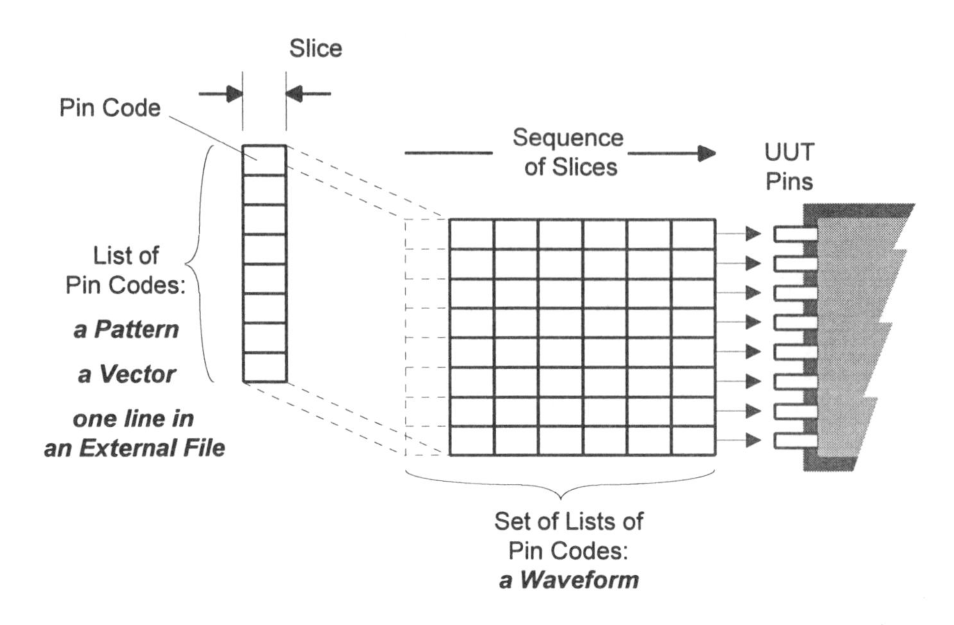

Figure 4-3. A Waveform described as a Set of Pin Code Lists

Here, we see that an entire waveform may be described as a set of pin code lists, in a two-dimensional, *data-sheet* sort of format. We will appreciate the significance of this format for describing waveforms better when we encounter the frame set, later in this chapter. This figure also illustrates some terminology extensions which we will encounter in the remainder of the text. The terms *pattern* and *vector* are used synonomously to indicate a *list of pin codes* for a particular slice of time. This is simply an artifact of the different types of users who deal with WAVES, including VHDL programmers, test engineers and designers, all of which have their own sets of terminology conventions. Now, having presented this conceptual view and introduced some terminology nuances, we next describe the relationship of a pin code to a frame in a waveform. Then, we present an example package developed to support the IEEE standard logic simulation environment.

Although we defer our detailed discussions of the frame set until later in this chapter, it is useful at this juncture to understand the relevance of the pin codes to this frame set. Each frame within a frame set has a one-character pin code associated with it. In this manner, the set of pin codes for each frame set are simply those used to define all the waveform shapes, for the frame set which is associated with a device test pin. Hence, the pin codes for a frame set define the wave shapes which will be used with whichever device pin or pins are associated with that frame set. In WAVES, the pin codes are a user-defined constant string. This is because the number and types of the pin codes necessary to describe a waveform depend on the complexities of the waveform, design, and testing environments. Instead of discussing the pin codes for an arbitrary environment, we have again chosen the IEEE standard logic simulation environment.

Now we begin to appreciate the value of the pin codes. As a pattern or vector for a slice of time, they define the waveform shapes we wish to apply in that slice of time. As a set of codes in a frame set, they define all the legal codes we may use for a particular pin. Hence, through this single, simple, shorthand notation we can associate our two principal dimensions of the waveform - time and waveform shape - with the test pins. Pin codes are the "glue" for the waveform building blocks.

Now we can get into more VHDL specifics. The pin codes package **Waves_1164_Pin_Codes**, developed by the USAF Rome Laboratory, is presented here as an example, and we have also included this package in Appendix B and on the companion CD-ROM (the name of the file is "**Wav_1164.Vhd**" located in the **WAV_PACK** directory). It is also reusable and is included in the library **WAVES_1164** to support the IEEE standard logic simulation environment. This package contains the constant declaration of a string that enumerates all of the legal 1164 pin codes that may appear in the WAVES external file, which serves to associate the pin codes with their actual slice starting times, as we shall explain in Chapter 5. The name of this constant is **Pin_Codes**, and we illustrate the code for this package in Figure 4-4.

```
PACKAGE WAVES_1164_pin_codes IS

-- Pin code declaration
  CONSTANT pin_codes : STRING := "X01ZWLH-";

END WAVES_1164_pin_codes;
```

Figure 4-4. WAVES_1164_Pin_Codes Package.

Note that, in using the WAVES 1164 convention, we have chosen to use the same set of pin codes as the logic value names we used in our Logic Value system (Section 4.1, above). This is not merely coincidence, but borne of a desire to have the pin codes imply something about the logic levels within the waveform shapes to which they refer. As we discussed in Chapter 3, our pin codes could have been any set of characters we chose, such as a, b, c, and d. However, there is value in making this association with logic levels, as we will see when we discuss the frame sets in the next section of this chapter. The pin codes string declaration of this example includes all of the legal IEEE standard 1164 codes except 'U' (uninitialized). The order of the elements of the pin codes string is not significant. Here, they have been declared in this order for consistency with IEEE STD 1164-1993. Basically, this declaration asserts that the pin codes in this string are the only legal ones for the simulation which may be using it. If we wished to implement some different codes, we would need to declare a different string. Remaining consistent with our use of the standard codes in our logic value system, we again omit the 'U' from the list of 1164 codes that are allowed as pin codes.

In ***summary of the pin codes,*** we see that the pin code is a simplified notation to represents a partial waveform (a frame) for an effective communication among different design and testing environments. As an example, we presented a package which supports the IEEE standard logic simulation environment, using the same standard logic codes as for the logic value system. Now, we will discuss the third reusable element of the WAVES dataset - the frame set.

4.3 Frames and Frame Sets

As we described in Chapter 3, the frames and frame sets are the most essential WAVES building block needed to implement a waveform within the WAVES-VHDL simulation environment. The purpose of the frame set is to define a set of frames which are sufficient to describe a certain signal for a particular pin. The frames in each frame set are indexed with the Pin Codes we described in Section 4.2, where each pin code refers to a particular frame. Hence, there are the same number of frames in each frame set as the number of pin codes. Each frame represents a waveform shape that is defined with the number of events and corresponding logic level transitions for each event. The logic value system we described in Section 4.1 is used to define the logic levels between transition events. Hence, we see that the frame set brings the shape, and logic level information together under the pin code indexing system.

Looking beyond the frame set itself, we need to provide the actual time at which each event occurs in order to define an unambiguous frame for a signal. For our purposes, we have chosen to provide a generic set of time identifiers. At the frame or frame set level, these times are not explicit values, but merely event time designators, such as t0, t1, and so forth. The actual event time values, with respect to the starting point of the frame, are specified in the frame set array, within an entity called a waveform generator procedure. This waveform generator procedure constructs a complete waveform, utilizing the WAVES building blocks described in this chapter, and it will be presented in Chapter 5. Our concern here remains simply with the frame set.

In this section, as an example of our frame set, we describe a WAVES package, named **Waves_1164_Frames**, that defines a set of basic waveform formats (shapes) that may be used to construct complete, complex waveform descriptions for all the input and output signals for a given Unit Under Test (UUT). This package provides a library of reusable waveform shapes (frame set formats) for our use and convenience when we are developing VHDL models compliant with the **IEEE Std_Logic Package**. We have included the waveform package **Waves_1164_Frames** in Appendix C and in the file called "**Wav_1164.Vhd**" located in the **WAV_PACK** directory on the companion CD-ROM. However, before we describe the frame set formats, we first define some basic terminology, to assist our transition of the conceptual waveform frame descriptions of Chapter 3 to the particular WAVES constructs we must actually use.

4.3.1 Terminology and Waveform Concepts Review

In this section, we first present some terminology and a waveform concepts review, to assist in making the transition from the waveform concepts of Chapter 3 to the actual WAVES constructs. Then, we introduce the package **Waves_1164_Frames**, developed by the USAF Rome Laboratory, which defines the frame set formats to support waveform construction. Now, we begin with the terminology.

In WAVES, a waveform is comprised of three basic elements: **pattern data** (the *truth table* data - which are the pin code lists for the time slices), **format** (the waveform shape), and **timing** (the edge transition points). These three elements combine to create the waveforms for data input to the UUT (the **drive** data) and the data expected on the output of the UUT (the **expect** or **compare** data). As a segment of a waveform, a frame must also contain these three elements: the format to specify the number and general nature of logic level transition events, the pattern data to index the specific shapes and logic values for the levels, and the timing to apply specific times to the events.

The pattern data, as we mentioned in Section 4.2, refers to the lists of legal pin codes that are valid, as they also must exist in a WAVES external file which provides the slice starting times. In our examples, the pattern data 'X', '0', '1', 'Z', 'W', 'L', 'H', or '-' occurs once for each WAVES slice (which also roughly may correspond to a tester cycle) for a given signal within the waveform. By itself, the pattern data means nothing to the UUT, because the frame data must also have a format, which is its shape, as well as logic values and event times. It is the job of the WAVES frame set descriptions to supply this additional information, which is necessary for the analysis system to properly "build" the data sheet-like waveforms. Then, in waveform generation, the pattern data can cause the actual logic values to be applied to the UUT in the chosen format(s) at the times specified in order to create waveform(s) at the UUT pins.

A WAVES-formatted signal is comprised of several WAVES elements that are built up to define a static definition of the waveform shape and associated timing. In this section we describe the **Waves_1164_Frames** package, which was developed to establish a basic set of waveform shapes that are readily available for our use. We preface our discussion with a quick review of the waveform concepts and building elements described earlier in Chapter 3.

The time segments associated with a waveform can be thought of as consecutive, advancing slices of a waveform. These segments of time are called WAVES slices and roughly correspond to test system "cycles". A tester **cycle** is one tester period and is measured from t0 (time zero) of one cycle to t0 of the next cycle, as we illustrate in the 2-slice waveform segment of Figure 4-5. Hence, in our context, *we will use cycle and slice synonymously*. In this figure, we also illustrate the source of the slice information - with the format (shape) of a slice of an individual signal coming from a frame and the complete slice format information for the waveform coming from a frame set array, as we described earlier.

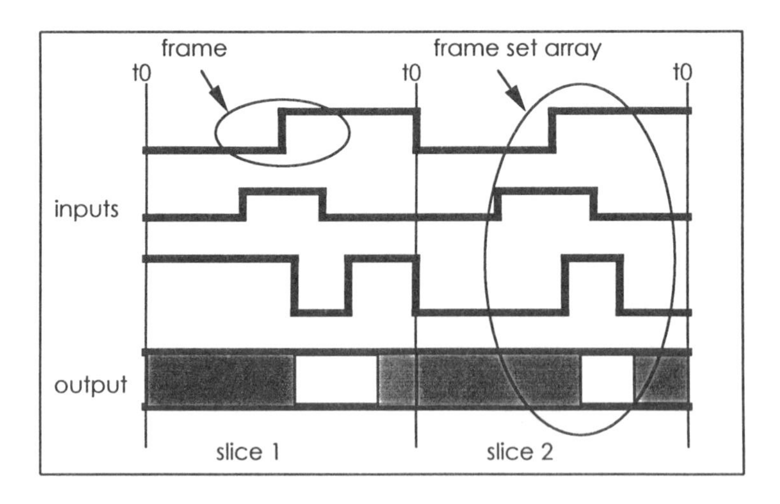

Figure 4-5. Sample Input/Output Waveforms for Two Consecutive WAVES Slices

Typically, each individual waveform may have multiple logic level edge transitions in one period or cycle. This list of edge transitions (WAVES events) on an individual signal of the waveform within a slice is defined as a **frame**. The set of frames for all possible legal pattern values (WAVES pin codes) that can be used on a signal (for one or more UUT test pins) is called a **frame set**. The functions that we describe in this section are used to define and establish a *common* set of the waveform shapes (WAVES frame sets). Later, in Chapter 5, we will show how these waveform shapes are combined to create a **frame set array** to describe the waveform signals on all the UUT inputs and outputs in the time slices. For the rest of this section we will refer to the pre-defined collection of waveform shapes, within the frame sets, as ***formats***. In other words, the shape within a frame is its format. Now with our terminology and waveform concepts review completed, we are ready to discuss implementation of the frame sets.

In order to verify the functionality and performance of a digital design, an analysis system must supply a set of input signals and know the expected outputs which will come from the device and know when the outputs are expected to come (in order to flag any errors). To support this analysis system, we have two types of frame set formats; *Drive format* for input signals and *Expected format* for the expected outputs. We introduce each format here, and describe them in detail later in this section.

First, the drive formats are designated as the waveform shapes that are associated with a UUT *input signal*. That is, what the signal does during the slice when the pin code is specified. The **Waves_1164_Frames** package contains a reasonable, generic set of drive formats for the 1164 logic values that may be used by anyone. The WAVES drive formats include: **non-return** (NR), **return high** (RH), **return low** (RL), **pulse high** (PH), **pulse high skew** (PHS), **pulse low** (PL), **pulse low skew** (PLS), and **surround by complement** (SC, also known as return to complement (RC)). The **Waves_1164_Frames** package contains a *function* which corresponds to each of these formats, which we shall detail later.

Second, for the expected format (output signals), we have two WAVES format functions to supply the analysis system with the information necessary to compare the actual device under test output with the expected output at a specified time. The compare is performed over a span of time by using a **window** compare or a **window skew** compare function. Again, we shall detail these formats a little later in this section.

These WAVES frame set formats provide great flexibility and may be used to build very complex waveforms. The package specification that defines these frame sets is shown in Figure 4-6. Here, we see the organization and sequence of the function statements we use to declare the formats we introduced earlier, and this figure is intended simply to illustrate that organization. First the context clauses provide visibility of the other packages, which include the pin codes (**WAVES_1164_Pin_Codes**) and logic value (**WAVES_1164_Logic_Value**) packages, as well as an interface package (**WAVES_Interface**), which simply provides the means to associate entities within the frame set array. Next, the frame set package is declared (PACKAGE **WAVES_1164_Frames** IS). Finally, all the individual frame sets are declared as function statements. In these frame set function statements, we see the time designators and a time unit designator (**t1** : TIME, for example), and that the function returns a frame set (RETURN **Frame_Set**).

```
-- Context Clauses
LIBRARY WAVES_1164;
USE WAVES_1164.WAVES_1164_pin_codes.ALL;
USE WAVES_1164.WAVES_1164_logic_value.ALL;
USE WAVES_1164.WAVES_interface.ALL;
PACKAGE WAVES_1164_frames IS
 --
 -- Declare functions that return Frame Sets.
 --
  -- Drive format function declarations

  FUNCTION non_return( t1 : TIME ) RETURN frame_set;
  FUNCTION return_low( t1, t2 : TIME ) RETURN frame_set;
  FUNCTION return_high( t1, t2 : TIME ) RETURN frame_set;
  FUNCTION surround_complement( t1, t2 : TIME )
                         RETURN frame_set;
  FUNCTION pulse_low( t1, t2 : TIME ) RETURN frame_set;
  FUNCTION pulse_low_skew( t0, t1, t2 : TIME )
                         RETURN frame_set;
  FUNCTION pulse_high( t1, t2 : TIME ) RETURN frame_set;
  FUNCTION pulse_high_skew( t0, t1, t2 : TIME )
                         RETURN frame_set;

  -- Expected data (compare) format function declarations

  FUNCTION window( t1, t2 : TIME ) RETURN frame_set;
  FUNCTION window_skew( t0, t1, t2 : TIME ) RETURN frame_set;

END WAVES_1164_frames;
```

Figure 4-6. WAVES_1164_frames Package declaration

Before proceeding, there are some nuances of the package illustrated in Figure 4-6, applicable to all our frame set discussions in this chapter, which we need to point out. First of all, in our WAVES 1164 convention, we are constrained to only two additional event times (t1 and t2), in addition to a shift or delay time for the skew cases (t0), in our frames. Hence, our function statements of Figure 4-6 contain only these three time designators. Please note that *this t0 is a shift time and not the t0 denoting the beginning of a slice*, which we discussed earlier. Also, these function statements are specific to a particular format (return low, pulse high, and so forth). Hence, we see that by implementing our frame set function statements in this manner, we *constrain any given frame set to only one shape or format.* As we shall soon explain, *the differences among the frames in the frame set are a function of logic level.* We point this out because our general waveform discussions of Chapter 3 implied that a frame set could contain many different frame shapes - and it can. Here, however, we are restricting our example to how we implement a frame set of one shape and varying logic levels, and leave the aspect of additional shape formats in frame sets to a later chapter on advanced topics. Although these two aspects - two-event frames and single-format frame sets - may appear restrictive, in practice they provide much of the flexibility we require. The balance between flexibility and standardization is always an issue for lively debate, and we shall not go into that here. The advantages of this paradigm will become apparent as we explain each function in this package in detail in subsequent sections.

In summary of our terminology and concepts overview, we presented the necessary information to assist in making the transition from the waveform concepts to the WAVES building block constructs. We also introduced the package specification of the **Waves_1164_Frames** to support the IEEE standard logic simulation environment. This package provides two types of the frame set, the drive and expected format, that can be used to create a complete WAVES waveform. Now, before presenting the WAVES implementation of each function, we need to discuss the data-dependent characteristics of the frame set format.

4.3.2 Data-Dependent Formats

The term *data-dependent format* simply implies that some additional source of data is required to complete the specification of some entity. As we illustrated in Figure 4-6, the functions in this package require additional timing specifications to define the frame set. With this additional timing specification, the waveform can be represented in a form that is entirely dependent upon the data in the pattern vectors - the datasheet representation we have mentioned earlier. As we shall illustrate later in this section, usually, for each individual signal cycle or slice, we are concerned with just one pair of timing edge transitions: the **leading** edge transition t1 and the **trailing** edge transition t2, with an occasional need for a t0 (shift time) for the skew cases. The formats used in our **Waves_1164_Frames** package, as we illustrated in Figure 4-6, conform to this convention. Since the waveform shapes are entirely dependent upon the data expressed in the pattern vectors (pin codes), we call them *data-dependent*. In our application, another advantage of this data-dependent format is that it permits us to specify values both before and after the particular time designators. In this context, our data-dependent formats always present pattern data at least between the t1 and t2 markers. However, the format selected also describes the logic level for the surrounding times (that is from t0 at the beginning of the cycle until t1 and from t2 until t0 at the end of the current cycle/beginning of the next cycle). It is a useful notation, as we shall see within the frame set formats of the ensuing sections.

Now, with understanding of these basic terminology and data dependent characteristics of the frame set in place, we can describe the two frame set formats; the drive format and expected data format (sometimes called compare format). We begin with the drive formats.

4.3.3 Drive Formats

In this section, and the next one, we describe some particular frame formats. Here, many of the concepts we have discussed earlier, as individual entities, begin to come together into a cohesive form for waveform description. We need to become completely comfortable with these frame formats before proceeding further, as they are the true foundation for WAVES. The concepts are simple and the VHDL notation is straightforward. By taking the time to appreciate the differences among the various formats here, we will find the more advanced material in subsequent chapters quite easy to grasp. We think of frame formats as the "alphabet" of WAVES, a necessary precursor to writing "sentences" later.

Now we may begin our drive formats presentation. As the name implies, the drive formats are designated as the waveform shapes we would include in a frame set that is associated with a UUT input signal. These formats are used to specify what the signal does during the time slice when a particular pin code is assigned to a UUT pin. The drive formats can be classified into two categories: ***compound*** and ***pulse*** formats. The compound formats category consist of four functions, which are **non-return** (NR), **return high** (RH), **return low** (RL), and **surround by complement** (SC). In section 4.3.3.1, we present these functions in detail. The pulse formats also consist of four functions, which are **pulse high** (PH), **pulse high skew** (PHS), **pulse low** (PL), **pulse low skew** (PLS). These pulse format functions will be presented in Section 4.3.3.2. We now begin our drive formats discussions with the compound formats.

4.3.3.1 Compound Formats

The compound formats are used to specify non-periodic or irregular partial signal segments in a slice or cycle. If a signal has a periodic nature, it can be described more efficiently with the second category, pulse formats, as we shall describe later. However, the compound formats are more general, and *any* segment of a waveform, even those exhibiting periodic characteristics, can be described by using the compound formats. As we mentioned above, **non-return** (NR), **return high** (RH), **return low** (RL), and **surround by complement** (SC) are the functions that support the compound formats. Next, we describe each of these function individually. In each case, we begin with a description of the format and include a graphical portrayal of its shape, then provide the actual VHDL code for the function.

Non-Return (NR): A typical output from a simulation program is in NR format, which is the simplest way to represent device behavior. The NR frame forces a shape data transition at t1 (only), and continues driving the data to the designated logic level until the next t1, ignoring the subsequent t2 and the next slice's t0. That is, at t0 the drive level is whatever the data of the previous cycle (slice) had been. At t1, it drives the logic level of the present cycle to the designated value, where it remains at least until t1 of the following cycle, ignoring both that cycle's t2 and the subsequent cycle's t0. A sample signal consisting of two consecutive cycles, showing the NR format, is given in Figure 4-7 and the NR frame set definition is shown subsequently in Figure 4-8. In this figure, and in all subsequent graphical format depictions, we use the following conventions: (1) the solid portion of the shape indicates that which is actually being specified or driven, (2) the shaded area indicates an undefined region, (3) a dashed portion (which does not appear in this particular figure) denotes a default value, due to the frame set definition, and (4) the "data = x" indicates the pattern data for the particular frame.

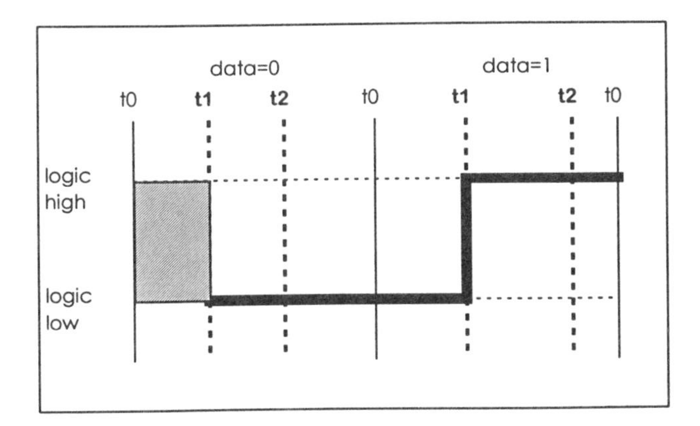

Figure 4-7. Non-Return Format

The definition of Figure 4-8 illustrates how we establish the relationship between the pin codes and the frames by using a function called **New_Frame_Set**. For example, the code **New_Frame_Set** ('1', **Frame_Event((Drive_1, Edge)))** defines that pin code '1' represents an occurrence of logic value "**Drive_1**" at **Event_Time** t1 in a frame. Now we can appreciate the use of standard logic codes as pin codes. The pin code implies the logic level for the portion of interest in the frame, according to the format. Hence, in our datasheet representation of an entire waveform, the pin codes actually represent the levels we wish to implement, and the pattern data in the datasheet provides an immediate picture of the nature of the waveform. The event in a frame is defined by the **Frame_Event** function. This function simply captures an event (i.e., occurrence of a logic value at time t1) in a frame. These functions, **New_Frame_Set** and **Frame_Event**, are defined in the **Waves_Interface** standard package. We discuss this package later, in Chapter 5.

```
FUNCTION non_return( t1 : TIME ) RETURN frame_set IS

    CONSTANT edge : event_time := etime( t1 );

  BEGIN
    RETURN
      new_frame_set( 'X', frame_event( (drive_X, edge) ) ) +
      new_frame_set( '0', frame_event( (drive_0, edge) ) ) +

     -- Here, pin code '1' represent
     --    an event (occurrence of drive_1 at time t1
      new_frame_set( '1', frame_event( (drive_1, edge) ) ) +
      new_frame_set( 'Z', frame_event( (drive_Z, edge) ) ) +
      new_frame_set( 'W', frame_event( (drive_W, edge) ) ) +
      new_frame_set( 'L', frame_event( (drive_L, edge) ) ) +
      new_frame_set( 'H', frame_event( (drive_H, edge) ) ) +
      new_frame_set( '-', frame_event );
  END non_return;
```

Figure 4-8. Non-Return Frame set declaration

Here, we note that a frame event with no parameters returns an empty frame. That is, when non-return is used with a pin code of "-" no events occur in the given slice.

In ***summary of our non-return format***, we see that it provides for driving the segment of interest in the frame, in this case the portion after t1 until the next cycle's t1, to any of the eight standard logic values.

Return High (RH): The RH frame, the second compound drive format, drives the level high from t0 to t1 and from t2 to the following t0. That is, the drive level is high at t0, transitions at t1, but only if the pattern data is low, and returns high at t2. A sample waveform showing the RH format is shown in Figure 4-9. The heavy dashed line represents the drive level present due to the frame set definition (in this case Return High) and not from the pattern data shown at the top of the figure. In other words, a "Return High" must be "high" to begin, and must subsequently return there.

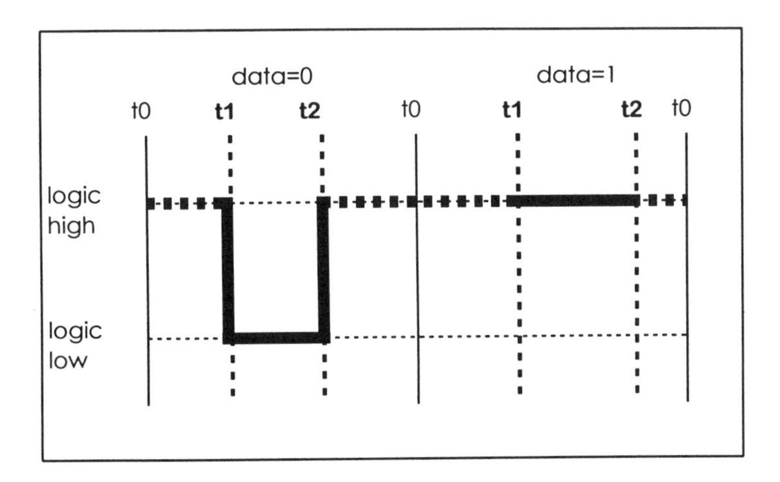

Figure 4-9. Return High Format

The RH frame set definition is shown in Figure 4-10. Here, a function **Frame_Elist** (frame event list) is used to capture multiple events in a frame, since the **Frame_Event** function only captures one event. For example, the code **New_Frame_Set('0', Frame_Elist(((Drive_0, Edge1), (Drive_1, Edge2))))** defines that the pin code '0' represents two events in a frame; occurrence of the logic value **Drive_0** at t1 followed by **Drive_1** at t2. This function also contains an error checking routine since an event at t1 must occur prior to the event at t2.

```
FUNCTION return_high( t1, t2 : TIME ) RETURN frame_set IS
    CONSTANT edge1 : event_time := etime( t1 );
    CONSTANT edge2 : event_time := etime( t2 );
  BEGIN
    -- Error checking routine
```

```
      ASSERT t1 < t2
      REPORT "Timing violation in Return_High frames." &
             "The inequality: T1 < T2 Must hold."
      SEVERITY FAILURE;
      RETURN
        new_frame_set( 'X', frame_elist( ((drive_X, edge1),
                                          (drive_1, edge2)) ) ) +
-- Pin code '0' represents two events in a frame; occurrence
-- of the logic value drive_0 at t1 followed by drive_1 at t2

        new_frame_set( '0', frame_elist( ((drive_0, edge1),
                                          (drive_1, edge2)) ) ) +
        new_frame_set( '1', frame_event( ( drive_1, edge1) ) ) +
        new_frame_set( 'Z', frame_elist( ((drive_Z, edge1),
                                          (drive_1, edge2)) ) ) +
        new_frame_set( 'W', frame_elist( ((drive_W, edge1),
                                          (drive_1, edge2)) ) ) +
        new_frame_set( 'L', frame_elist( ((drive_L, edge1),
                                          (drive_1, edge2)) ) ) +
        new_frame_set( 'H', frame_elist( ((drive_H, edge1),
                                          (drive_1, edge2)) ) ) +
        new_frame_set( '-', frame_event( ( drive_1, edge2 ) ) );
    END return_high;
```

Figure 4-10. Return High Frame set declaration

In ***summary of our Return High format***, we see that it provides for beginning at a logical high, then driving the segment of interest in the frame, in this case the portion between t1 and t2, to any of the eight standard logic values, and finally returning to a logical high.

Return Low (RL) The RL frame, the third compound format, is basically the complement of the RH format. It drives the level low from t0 to t1 and from t2 to the following t0. That is, the drive level is low at t0, transitions at t1 but only if the pattern data is high, and returns low at t2. A sample waveform showing the RL format is shown in Figure 4-11. Again, the heavy dashed line represents the drive level present due to the frame set definition and not from the pattern data.

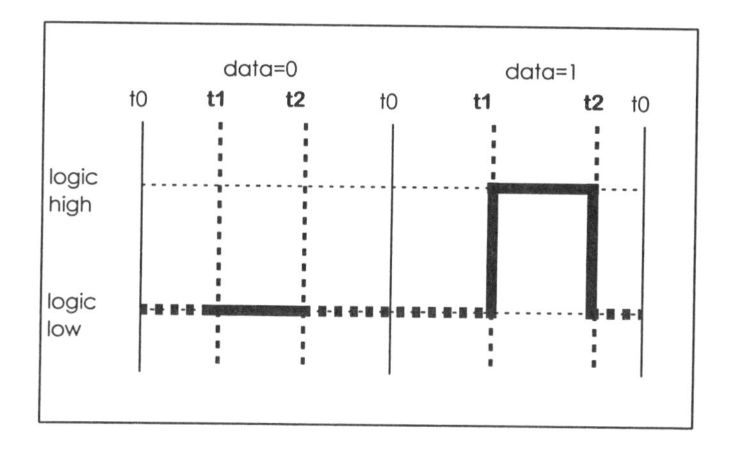

Figure 4-11. Return Low Format

The RL frame set definition is shown in Figure 4-12. This frame set definition is very similar to the RH described above, except its default logic level is "low."

```
FUNCTION return_low( t1, t2 : TIME ) RETURN frame_set IS
    CONSTANT edge1 : event_time := etime( t1 );
    CONSTANT edge2 : event_time := etime( t2 );
  BEGIN
    -- Error checking routine
    ASSERT t1 < t2
    REPORT "Timing violation in Return_Low frames." &
           "The inequality : T1 < T2 Must hold."
    SEVERITY FAILURE;
    RETURN
      new_frame_set( 'X', frame_elist( ((drive_X, edge1),
                                        (drive_0, edge2)) ) ) +
      new_frame_set( '0', frame_event( ( drive_0, edge1)  ) ) +

-- Pin code '1' represents two events in a frame; occurrence of
-- the logic value drive_1 at t1 followed by drive_0 at t2

      new_frame_set( '1', frame_elist( ((drive_1, edge1),
                                        (drive_0, edge2)) ) ) +
      new_frame_set( 'Z', frame_elist( ((drive_Z, edge1),
                                        (drive_0, edge2)) ) ) +
      new_frame_set( 'W', frame_elist( ((drive_W, edge1),
                                        (drive_0, edge2)) ) ) +
      new_frame_set( 'L', frame_elist( ((drive_L, edge1),
                                        (drive_0, edge2)) ) ) +
```

```
    new_frame_set( 'H', frame_elist( ((drive_H, edge1),
                                       (drive_0, edge2)) ) ) +
    new_frame_set( '-', frame_event( ( drive_0, edge2 ) ) );
END return_low;
```

Figure 4-12. Return Low Frame set declaration

In ***summary of our Return Low format***, we see that it provides for beginning at a logical low, then driving the segment of interest in the frame, in this case the portion between t1 and t2, to any of the eight standard logic values, and finally returning to a logical low.

Surround By Complement (SC) The SC frame, the last of the compound drive formats, drives the complement of the pattern data from t0 to t1 and from t2 to the following t0. That is, the drive level is at the complement of the pattern data at the beginning of the cycle, transitions to the pattern data level at t1, and returns to the complement at t2. Hence the name, "surround-by-complement." A sample waveform showing the SC format is shown in Figure 4-13. As usual, the heavy dashed line represents the drive level present due to the frame set definition and not from the pattern data. Note that in this specific example, when the complement of a given frame in one slice happens to be different than the complement value in the frame of the next slice, a transition occurs at t0 of the second slice, by default. This is basically an implied transition. It occurs as a function of the drive values we have chosen.

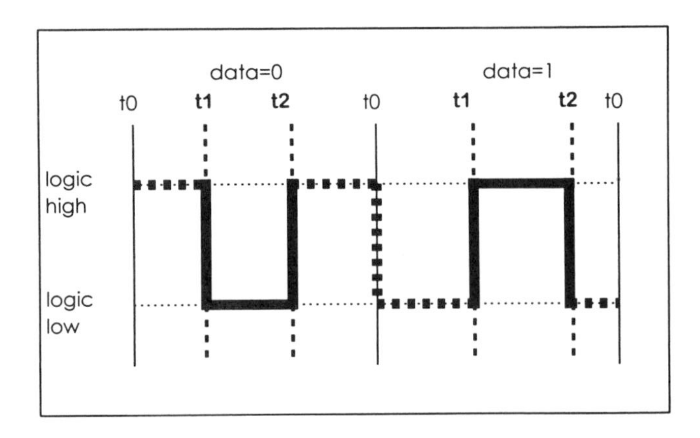

Figure 4-13. Surround by Complement Format

The SC frame set definition is shown in Figure 4-14. Again, it is quite similar to the RH and RL definitions given previously.

```
FUNCTION surround_complement( t1, t2 : TIME ) RETURN frame_set IS

    CONSTANT edge1  : event_time := etime( t1 );
    CONSTANT edge2  : event_time := etime( t2 );

  BEGIN

-- Error checking routine
    ASSERT t1 < t2
    REPORT "Timing violation in surround_Complement frames." &
          "The inequality: T1 < T2 Must hold."
    SEVERITY FAILURE;
    RETURN
      new_frame_set( 'X', frame_event( ( drive_X, edge1)   ) ) +
      new_frame_set( 'O', frame_elist( ((drive_0, edge1),
                                         (drive_1, edge2)) ) ) +

-- Pin code '1' represents two events in a frame; occurrence of the
-- logic value drive_1 at t1 followed by drive_0 at t2
      new_frame_set( '1', frame_elist( ((drive_1, edge1),
                                         (drive_0, edge2)) ) ) +
      new_frame_set( 'Z', frame_event( ( drive_Z, edge1)   ) ) +
      new_frame_set( 'W', frame_event( ( drive_W, edge1)   ) ) +
      new_frame_set( 'L', frame_elist( ((drive_L, edge1),
                                         (drive_H, edge2)) ) ) +
      new_frame_set( 'H', frame_elist( ((drive_H, edge1),
                                         (drive_L, edge2)) ) ) +
      new_frame_set( '-', frame_event );
  END surround_complement;
```

Figure 4-14. Surround by Complement Frame Set declaration

In ***summary of our Surround by Complement format***, we see that it provides for beginning at the complement of the pattern data, then driving the segment of interest in the frame, in this case the portion between t1 and t2, to any of the eight standard logic values, and finally returning to the complement.

In ***summary of our Compound Formats***, we have presented four functions that are used to specify non-periodic or irregular partial waveform in a slice for input signals. These functions are reusable and included in the library **WAVES_1164** to facilitate generation of the WAVES dataset. Now, let us consider the second drive formats, the pulse formats, to handle periodic nature of input signals in a waveform.

4.3.3.2 Pulse Formats

The pulse formats, the second category of the drive formats, are used to specify periodic or regular waveforms for input signals. If a signal has a periodic nature in a waveform (for example, a clock signal), it can be described more efficiently with the pulse formats. As such, pulse formats can drive a fixed waveform to the UUT, which is useful for supplying clock signals to it. There are two classes of pulse formats; **Pulse Low** and **Pulse High**. In addition, these frame set formats have two separate instantiations. The first instantiation (**Pulse_Low** and **Pulse_High**) allow for data pulses only within one WAVES slice (t0, which implies a time shift, is always set to 0 and t1 and t2 occur in the same slice). The second instantiation (**Pulse_Low_Skew** and **Pulse_High_Skew**) are simple time-shifted pulses. This allows the data pulse to be present across WAVES slice boundaries (t2 may not be in the same slice as t1, because the entire event time sequence is shifted by t0). In the following descriptions and examples we will detail this concept.

Pulse Low/Pulse Low Skew (PL/PLS). If the vector pattern data contains either an 'L' or a '0,' this format will drive to 0 on the first edge (t1), and stay at 0 until the second edge (t2). That is, it will transition from high-to-low at t1 and from low-to-high at t2. Note that for the PLS format the entire waveform is shifted into the next cycle by an amount depicted by t0, so that t2 occurs in a subsequent slice (cycle). The signal will stay high for all other values of vector pattern data. Sample signals illustrating the PL format are shown in Figure 4-15, and signals showing the PLS format are given in Figure 4-17. Here, we designate the t0 shift time as t0' to distinguish it from the t0 slice beginning time. We didn't invent this redundancy, we're just trying to live with it. The PL frame set definition is shown in Figure 4-16 and the PLS frame set definition is shown in Figure 4-18.

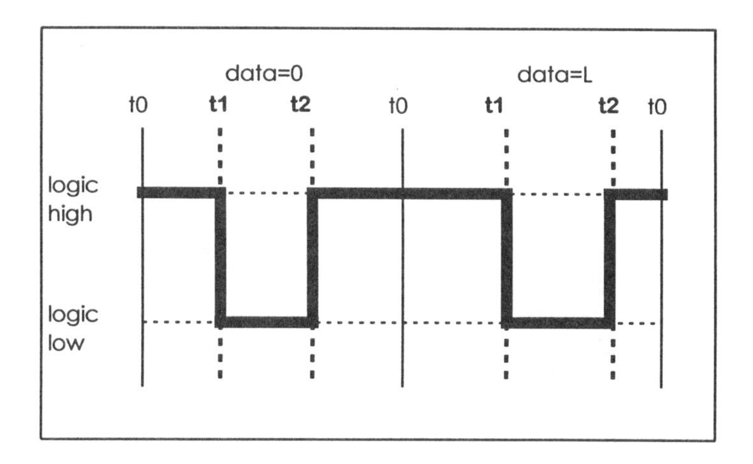

Figure 4-15. Pulse Low Format

```
FUNCTION pulse_low( t1, t2 : TIME ) RETURN frame_set IS

    CONSTANT edge0 : event_time := etime( 0 ns );
    CONSTANT edge1 : event_time := etime( t1 );
    CONSTANT edge2 : event_time := etime( t2 );

  BEGIN

-- Error checking routine
    ASSERT t1 < t2
    REPORT "Timing violation in Pulse_Low frames." &
           "The inequality: T1 < T2 Must hold."
    SEVERITY FAILURE;
    RETURN
      new_frame_set( 'X', frame_event ) +

      --  pin code '0' represents a frame containing a low pulse
      new_frame_set( '0', frame_elist( ((drive_1, edge0),
                                        (drive_0, edge1),
                                        (drive_1, edge2)) ) ) +
      new_frame_set( '1', frame_event( ( drive_1, edge0)  ) ) +
      new_frame_set( 'Z', frame_event ) +
      new_frame_set( 'W', frame_event ) +
      new_frame_set( 'L', frame_elist( ((drive_H, edge0),
                                        (drive_L, edge1),
                                        (drive_H, edge2)) ) ) +
      new_frame_set( 'H', frame_event( ( drive_H, edge0)  ) ) +
      new_frame_set( '-', frame_event );
  END pulse_low;
```

Figure 4-16. Pulse Low frame set declaration

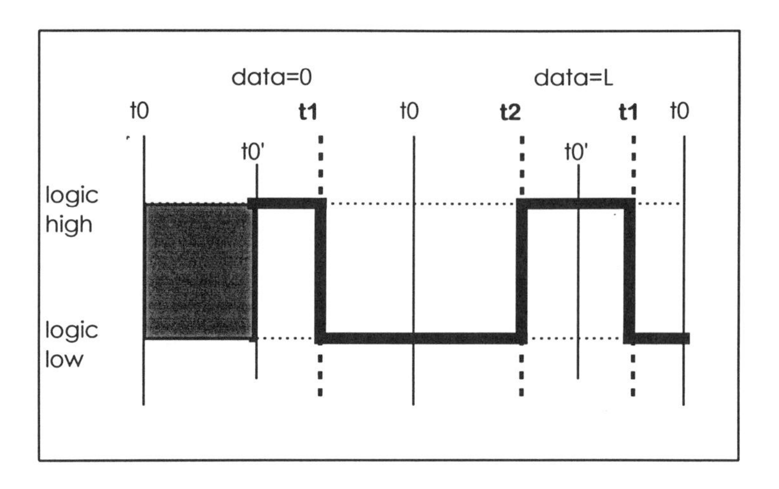

Figure 4-17. Pulse Low Skew Format

```
FUNCTION pulse_low_skew( t0, t1, t2 : TIME ) RETURN frame_set IS

    CONSTANT edge0 : event_time := etime( t0 );
    CONSTANT edge1 : event_time := etime( t1 );
    CONSTANT edge2 : event_time := etime( t2 );

  BEGIN

-- Error checking routine
    ASSERT t0 < t1 AND t1 < t2
    REPORT "Timing violation in Pulse_Low frames." &
           "The inequality: T0 < T1 < T2 Must hold."
    SEVERITY FAILURE;
    RETURN
      new_frame_set( 'X', frame_event ) +

      --  pin code '0' represents a frame that contain a low pulse
      --   In this function, t2 exist outside the current slice.
      new_frame_set( '0', frame_elist( ((drive_1, edge0),
                                        (drive_0, edge1),
                                        (drive_1, edge2)) ) ) +
      new_frame_set( '1', frame_event( ( drive_1, edge0)  ) ) +
      new_frame_set( 'Z', frame_event ) +
      new_frame_set( 'W', frame_event ) +
      new_frame_set( 'L', frame_elist( ((drive_H, edge0),
                                        (drive_L, edge1),
                                        (drive_H, edge2)) ) ) +
```

```
      new_frame_set( 'H', frame_event( ( drive_H, edge0)  ) ) +
      new_frame_set( '-', frame_event );
   END pulse_low_skew;
```

Figure 4-18. Pulse Low Skew Frame set declaration

Pulse High/Pulse High Skew (PH/PHS). These functions make up the second class of the pulse drive formats. If the vector pattern data contains either an 'H' or a '1' this format will drive to 1 on the first edge (t1), and stay at a 1 until the second edge (t2). That is, it will transition low-to-high at t1 and high-to-low at t2. Note that (similar to the PLS format described previously) for the PHS format the entire waveform is shifted into the next cycle by an amount depicted by t0, so t2 occurs in a subsequent slice (cycle). The signal will stay low for all other values of vector pattern data. Sample signals showing the PH format are shown in Figure 4-19 and those showing the PHS format appear in Figure 4-21. Again, we designate the t0 shift time as t0' to distinguish it from the t0 slice beginning time. The PH frame set definition is given in Figure 4-20 and the PHS frame set definition is shown in Figure 4-22.

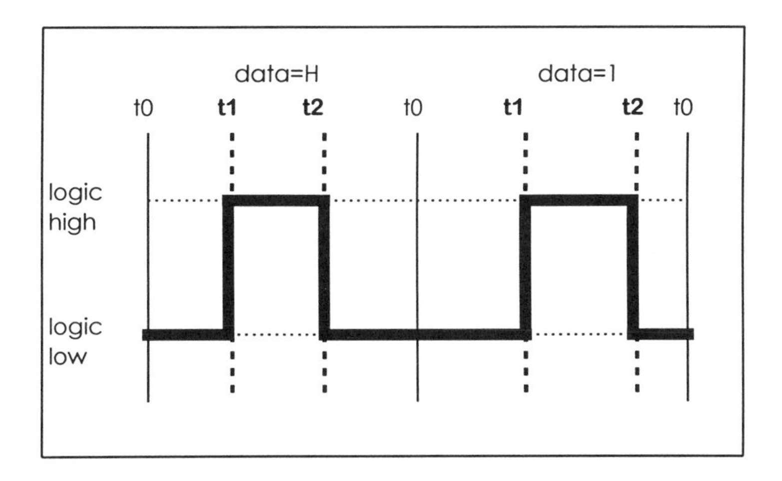

Figure 4-19. Pulse High Format

```
FUNCTION pulse_high( t1, t2 : TIME ) RETURN frame_set IS

   CONSTANT edge0 : event_time := etime( 0 ns );
   CONSTANT edge1 : event_time := etime( t1 );
   CONSTANT edge2 : event_time := etime( t2 );
```

```
  BEGIN

-- Error checking routine
    ASSERT t1 < t2
    REPORT "Timing violation in Pulse_High frames." &
           "The inequality: T1 < T2 Must hold."
    SEVERITY FAILURE;
    RETURN
      new_frame_set( 'X', frame_event ) +
      new_frame_set( '0', frame_event( ( drive_0, edge0)  ) ) +

     -- pin code '1' represents a frame that contain a high pulse
     new_frame_set( '1', frame_elist( ((drive_0, edge0),
                                       (drive_1, edge1),
                                       (drive_0, edge2)) ) ) +
      new_frame_set( 'Z', frame_event ) +
      new_frame_set( 'W', frame_event ) +
      new_frame_set( 'L', frame_event( ( drive_L, edge0)  ) ) +
      new_frame_set( 'H', frame_elist( ((drive_L, edge0),
                                       (drive_H, edge1),
                                       (drive_L, edge2)) ) ) +
      new_frame_set( '-', frame_event );
  END pulse_high;
```

Figure 4-20. Pulse High frame set declaration (first instantiation)

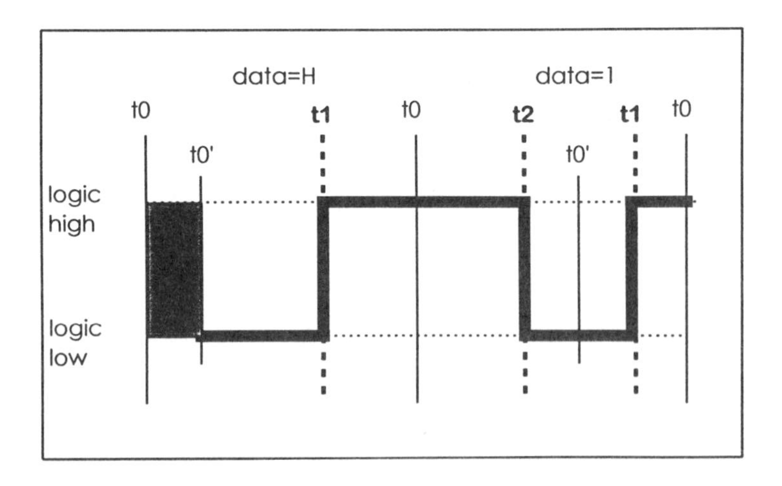

Figure 4-21. Pulse High Skew Waveform

```
FUNCTION pulse_high_skew( t0, t1, t2 : TIME ) RETURN frame_set IS

  CONSTANT edge0 : event_time := etime( t0 );
  CONSTANT edge1 : event_time := etime( t1 );
  CONSTANT edge2 : event_time := etime( t2 );

BEGIN
 -- Error checking routine
  ASSERT t0 < t1 AND t1 < t2
  REPORT "Timing violation in Pulse_High frames." &
         "The inequality: T0 < T1 < T2 Must hold."
  SEVERITY FAILURE;
  RETURN
    new_frame_set( 'X', frame_event ) +
    new_frame_set( '0', frame_event( ( drive_0, edge0)  ) ) +

     --  pin code '0' represents a frame that contain a low pulse
     --   In this function, t2 exist outside the current slice.
    new_frame_set( '1', frame_elist( ((drive_0, edge0),
                                      (drive_1, edge1),
                                      (drive_0, edge2)) ) ) +
    new_frame_set( 'Z', frame_event ) +
    new_frame_set( 'W', frame_event ) +
    new_frame_set( 'L', frame_event( ( drive_L, edge0)  ) ) +
    new_frame_set( 'H', frame_elist( ((drive_L, edge0),
                                      (drive_H, edge1),
                                      (drive_L, edge2)) ) ) +
    new_frame_set( '-', frame_event );
END pulse_high_skew;
```

Figure 4-22. Pulse High frame set declaration (second instantiation)

In ***summary of our pulse formats***, we have presented four functions in two classes that support the pulse formats which are utilized to define the frame sets for periodic input signals, with or without a selectable time delay or *skew*. These functions are reusable and are included in the library **WAVES_1164** to facilitate generation of the WAVES dataset.

In ***summary of our drive formats***, we can see that the drive formats can be classified into two categories: Compound and Pulse formats. Each category contains four functions and we utilize these functions to define the frame sets that can be used to construct a waveform specific to input signals (define the stimulus for a UUT). Next, we present frame sets defined to handle the output signals of the UUT - *Compare* formats.

4.3.4 Compare Formats

In the previous section, we described eight functions in the drive format used to generate the frame sets for input signals. Here, we present two WAVES format functions to supply the analysis system with the information necessary to compare the actual UUT output with the expected output at a specified time. The compare is performed over a span of time by using a **window** compare or a **window skew** compare function. Now, we will consider each format.

The **Window** format starts the valid data comparison at the t1 edge and disables the data comparison at the subsequent t2 edge. That is, the data comparison is true for the entire duration from t1 to t2. If the expected data level is not detected or changes to an invalid state at any time during the comparison interval, a **fail** is declared. An example of this Window format, where the data expected on the output of the UUT is low, high, and then midband (anywhere between the asserted levels), is shown in Figure 4-23. The Window frame set definition follows in Figure 4-24.

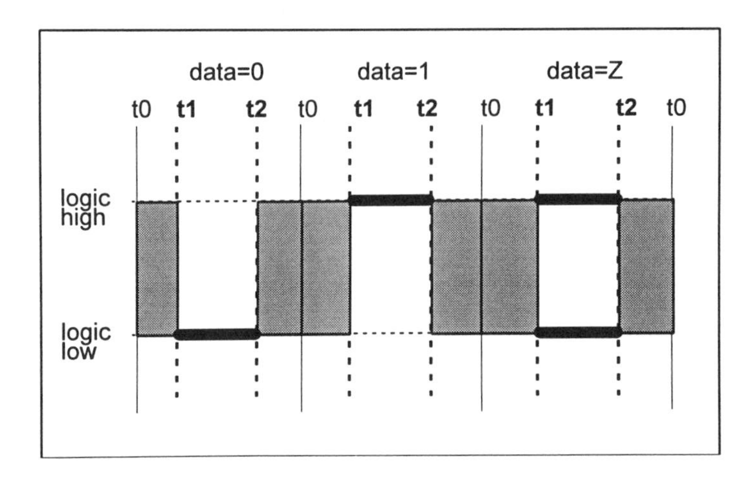

Figure 4-23. Window format

```
FUNCTION window( t1, t2 : TIME ) RETURN frame_set IS

    CONSTANT edge0 : event_time := etime( 0 ns );
    CONSTANT edge1 : event_time := etime( t1 );
    CONSTANT edge2 : event_time := etime( t2 );

  BEGIN
```

```
-- Error checking routine
    ASSERT t1 < t2
    REPORT "Timing violation in Window frames." &
           "The inequality: T1 < T2 Must hold."
    SEVERITY FAILURE;
    RETURN
      new_frame_set( 'X', frame_elist( ((dont_care, edge0),
                                          (sense_X, edge1),
                                          (dont_care, edge2)))) +

      --  pin code '0' represents that expected data must be
      --   logic low (i.e., sense_0) between t1 and t2

      new_frame_set( '0', frame_elist( ((dont_care, edge0),
                                          (sense_0, edge1),
                                          (dont_care, edge2)))) +
      new_frame_set( '1', frame_elist( ((dont_care, edge0),
                                           sense_1, edge1),
                                          (dont_care, edge2)))) +
      new_frame_set( 'Z', frame_elist( ((dont_care, edge0),
                                          (sense_Z, edge1),
                                          (dont_care, edge2)))) +
      new_frame_set( 'W', frame_elist( ((dont_care, edge0),
                                          (sense_W, edge1),
                                          (dont_care, edge2)))) +
      new_frame_set( 'L', frame_elist( ((dont_care, edge0),
                                          (sense_L, edge1),
                                          (dont_care, edge2)))) +
      new_frame_set( 'H', frame_elist( ((dont_care, edge0),
                                          (sense_H, edge1),
                                          (dont_care, edge2)))) +
     new_frame_set( '-', frame_event( ( dont_care, edge0)   ) );
  END window;
```

Figure 4-24. Window frame set declaration

The **Window Skew** format is similar to the Window format except that the event times are shifted or delayed by a particular amount, t0, much like the Pulse High Skew (PHS) and Pulse Low Skew (PLS) formats discussed in Section 4.3.3.2. An example waveform showing Window Skew formats where the data expected on the output of the UUT is low, high, and then midband, is shown in Figure 4-25. The Window Skew frame set definition is shown in Figure 4-26.

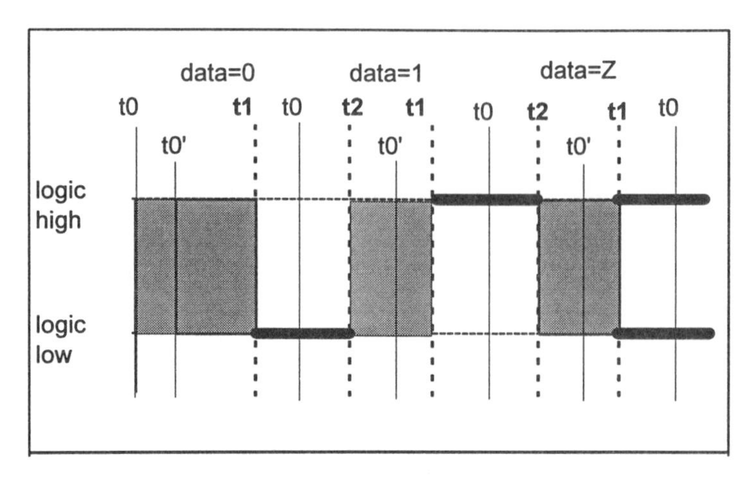

Figure 4-25. Window Skew format

```
FUNCTION window_skew( t0, t1, t2 : TIME ) RETURN frame_set IS

  CONSTANT edge0 : event_time := etime( t0 );
  CONSTANT edge1 : event_time := etime( t1 );
  CONSTANT edge2 : event_time := etime( t2 );

BEGIN

-- Error checking routine
  ASSERT t0 < t1 AND t1 < t2
  REPORT "Timing violation in Window frames." &
         "The inequality: T0 < T1 < T2 Must hold."
  SEVERITY FAILURE;
  RETURN
    new_frame_set( 'X', frame_elist( ((dont_care, edge0),
                                      (sense_X, edge1),
                                      (dont_care, edge2)))) +

    -- pin code '0' represents that expected data must be
    --  logic low (i.e., sense_0) between t1 and t2
    -- In this function, t2 exist outside the current slice

    new_frame_set( '0', frame_elist( ((dont_care, edge0),
                                      (sense_0, edge1),
                                      (dont_care, edge2)))) +
    new_frame_set( '1', frame_elist( ((dont_care, edge0),
                                      (sense_1, edge1),
                                      (dont_care, edge2)))) +
```

```
      new_frame_set( 'Z', frame_elist( ((dont_care, edge0),
                                        (sense_Z, edge1),
                                        (dont_care, edge2)))) +
      new_frame_set( 'W', frame_elist( ((dont_care, edge0),
                                        (sense_W, edge1),
                                        (dont_care, edge2)))) +
      new_frame_set( 'L', frame_elist( ((dont_care, edge0),
                                        (sense_L, edge1),
                                        (dont_care, edge2)))) +
      new_frame_set( 'H', frame_elist( ((dont_care, edge0),
                                        (sense_H, edge1),
                                        (dont_care, edge2)))) +
      new_frame_set( '-', frame_event( ( dont_care, edge0)) );
  END window_skew;
```

Figure 4-26. Window Skew Frame set declaration

As stated above, the compare window for valid data starts at time t1 and closes at time t2. Many analysis systems also have the capability of sampling data at a single point in time with what is referred to as an **edge strobe.** Using the window formats described above, it is possible to mimic an edge strobe by setting the t1 and t2 times to be as close to one another as possible, constrained only by the timing resolution of the analysis system.

In ***summary of our compare format***, we presented two WAVES format functions, which generate information (frame sets) necessary to compare the actual unit under test output with the expected output at a specified time. These functions are reusable and included in the library **WAVES_1164** to facilitate generation of the WAVES dataset.

In ***summary of our frames and frame set*s**, we can see that the frame set defines a set of frames for all possible legal pattern values (WAVES pin codes) that can be used on a signal. This frame set is an essential WAVES building block for constructing a waveform. In particular, we presented a recommended WAVES package, **Waves_1164_Frames**, which contains ten reusable functions that generate necessary frame sets for the IEEE standard logic simulation environment.

In ***summary of our chapter on WAVES Concepts***, we have illustrated the techniques for constructing three particular reusable elements of the WAVES dataset. Here, we described the construction process for the logic value system, the pin codes, and the frames and frame sets. In particular, the WAVES frames and frame sets described in this chapter provide great flexibility and can be used to build very complex waveforms. Next, in Chapter 5, we will present an overview of the design and test environment to illustrate the role of the WAVES dataset in this environment.

CHAPTER 5. THE WAVES DATASET

Putting it all together

As with most building-block techniques, once we have the elements, the next step is to put them together. In this chapter, we take the reusable WAVES elements we discussed in Chapter 4, and describe how to combine them with other device-dependent elements to realize the desired WAVES capability and to support our design or testing process. This process is called *constructing the WAVES dataset.* A WAVES dataset is the logical collection of data required to specify one or more waveforms completely. Here we go beyond the simple instruction of how to interface elements, and also describe the *relationships* we need to achieve among the elements. Our goal is to provide not only the techniques, but also to ensure that we obtain a solid understanding, which is essential for effectively applying the techniques to diverse design and testing problems.

We begin with an overview of a typical design and testing environment, and explain how the WAVES dataset is used in this environment. It is important to understand this *role* of the WAVES dataset, to appreciate the dataset construction process. After the role of the WAVES dataset is explained, we present the WAVES dataset structure which consists of three principal elements: the WAVES files, the external file, and a header file. Here, we describe the function of each of these three WAVES dataset elements in constructing a waveform. Finally, we present the WAVES Dataset Development Procedure, concentrating upon the analysis order of each WAVES dataset and the visibility of each package in the WAVES file. In this manner, we demonstrate how the WAVES building blocks, described in Chapter 4, are used to construct the waveforms. We now begin with our overview of the design and testing environment.

5.1 Overview of the Design and Testing Environment

In Figure 5-1 we illustrate, from an overview perspective, a typical design and test environment and how the WAVES dataset is utilized in this type of environment. During a design or testing cycle, WAVES generates a waveform to stimulate a UUT and to provide the expected response of the UUT to the WAVES-VHDL testbench. In this sense, the WAVES dataset captures all necessary testing requirements (test specifications) for the UUT and provide these data during simulation. Typically, the testbench instantiates a waveform generator, which resides in the WAVES dataset, as well as a UUT (VHDL model) to carry out the appropriate test. The testbench also contains monitor processes to compare the expected response to the actual response of the UUT.

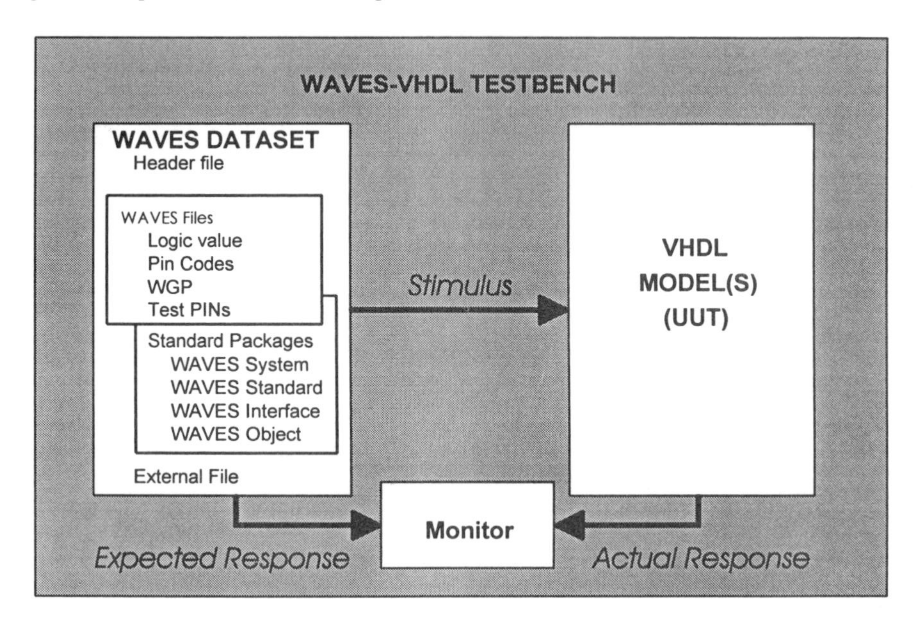

Figure 5-1. Overview of the Design or Testing Environment

Now, we see the role and function of the WAVES dataset in the design or test environment and can begin to understand the purpose of each element that makes up the WAVES dataset. We will present the WAVES-VHDL testbench generation in Chapter 6. In this chapter, our focus remains with the WAVES dataset.

5.2 The WAVES Dataset Structure

According to the WAVES LRM (IEEE STD 1029.1-1991), a WAVES dataset is the logical collection of data required to completely specify one or more waveforms. In general, it can be viewed as a collection of files which contains waveform information, such as the stimulus and expected response, and which specifies how to generate corresponding waveforms. In this sense, all the waveform building blocks, discussed in Chapter 3 and 4, and some additional information, are captured in a WAVES dataset. Now let's begin to explore an organizational view of the WAVES dataset.

Our WAVES dataset consists of *three elements*: (1) one or more WAVES files supported by WAVES standard packages, (2) zero or more external files of ancillary data, and (3) a header file, as we illustrate graphically in Figure 5-2. In this figure, the elements that may or may not exist are denoted by perforated boxes. Here, we only present a brief description of each element for an overview, and provide more detailed explanations of each element, and their integration, in later sections of this chapter. First, the *WAVES files* contain the WAVES specifications and most of the waveform building blocks discussed in Chapter 3 and Chapter 4. We present the nature and construction of the WAVES files in Section 5.3. Second, the *external files* contain a format (pin codes) that may be used by the WAVES files to produce a WAVES waveform. The external files are described further in Section 5.4. Third, the *header file* specifies how the dataset is to be assembled from the WAVES files and the external files. This file also provides some useful WAVES dataset identification information to us as WAVES users. The header file is described in detail in Section 5.5. We conclude this chapter with a WAVES dataset development procedure in Section 5.6.

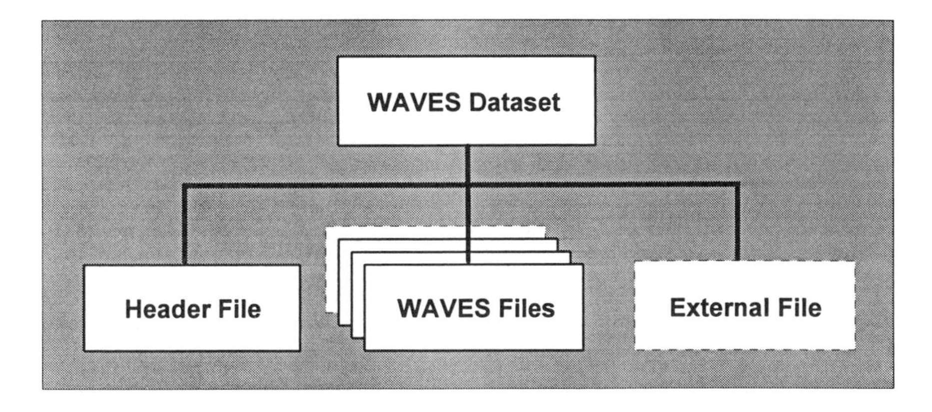

Figure 5-2. The WAVES Dataset

Having outlined an organizational view of the WAVES dataset, we now present a simple overview of how this WAVES dataset is established in a WAVES-VHDL simulation environment, using the strongly-typed and library-supported paradigms of VHDL. In a simulation environment, the WAVES dataset is analyzed or compiled into various libraries. In particular, the WAVES files may be analyzed into more than one library. Figure 5-3 shows an example library structure that may support the simulation. These libraries are necessary to contain the WAVES dataset and VHDL model, and to promote the interface between the WAVES and VHDL simulation environments. For example, some libraries such as the STD and **WAVES_STD** contain basic definitions necessary to support VHDL modeling and the WAVES dataset, and the WAVES dataset may be compiled into the **WAVES_ 1164** and Work libraries. At this point, we need not worry about the content and nature of these libraries. We should, however, realize that multiple libraries are needed to effectively support the WAVES-VHDL simulation environments. Hence, the examples we present in the following sections will contain some references to these libraries. The purpose and additional contents of these libraries will be presented in later sections and chapters as appropriate.

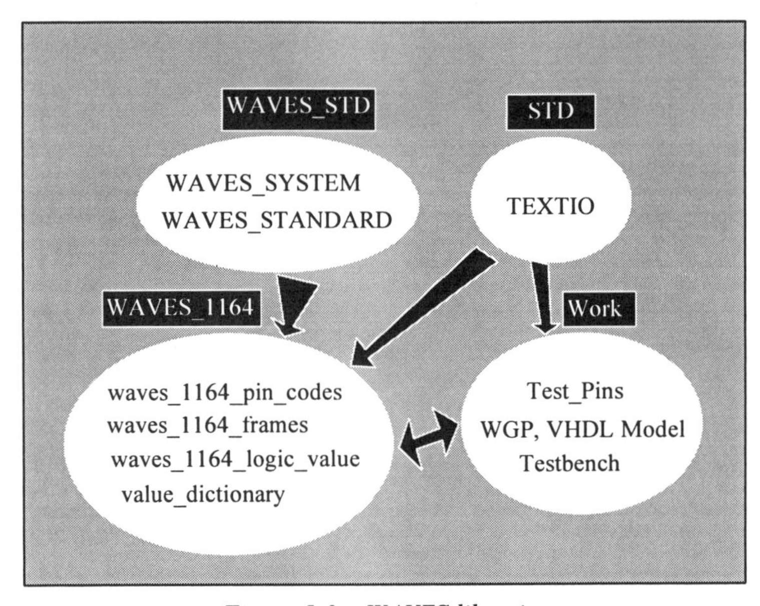

Figure 5-3. WAVES libraries

To summarize our dataset structure, the important points to remember here are that the WAVES dataset contains many packages and building blocks and that these packages and building blocks are organized into three principal elements: WAVES files, external files, and a header file. Also, some packages in the dataset are reusable and they are included (pre-compiled) in the **WAVES_1164** library. On the other hand, some packages are UUT dependent and those are compiled into the Work library. Now, having established these file and library relationships, we will discuss each of the three elements of the WAVES dataset individually.

5.3 WAVES Files

In this section, we describe our *first principal WAVES dataset element* - the WAVES Files. As we described in Chapters 3 and 4, a waveform can be specified by using many building blocks. Typically, each WAVES file contains a package that defines one or more of the waveform building blocks, which implies that the terms "WAVES files" and "WAVES packages" are interchangeable. In this section, we present the construction mechanism and WAVES codes for each building block, all of which are utilized to construct a complete waveform.

Through the WAVES files, we can implement a **Waveform Generator Procedure (WGP)** which generates slices of the waveform utilizing the WAVES frame and frame set building blocks and the pin codes. These slices of the waveform are then applied to the UUT during simulation. In order to completely define the WGP, the collection of WAVES files (packages) must provide certain required declarations and must conform to a package visibility restriction of VHDL. All the required declarations must appear within a package declaration, as this is a requirement specified in paragraph 3.4 of the WAVES Language Reference Manual (LRM) (IEEE STD 1029.1-1991). To handle these declarations, WAVES makes full use of the concept of libraries and packages embodied in VHDL.

Now, having previewed the WAVES file and package relationships, we can discuss each package. The packages, excluding the standard WAVES packages, which we need to describe using the WAVES files are: (1) a logic value system consisting of the Logic Values and a Value Dictionary, (2) Pin Codes, (3) Test Pins, (4) Frames and Frame Sets, and (5) a Waveform Generator Procedure (WGP). We have already described the logic value system, pin codes, frames, and frame sets in Chapter 4. These packages are reusable and included in the library **WAVES_1164**. In this section, we will explain WAVES packages that are not included in the **WAVES_1164** library. Typically, these packages are device specific (not reusable), but *utilize* the reusable building blocks in the **WAVES_1164** library.

First, in Section 5.3.1, we describe the test pins package that defines an interface for testing a UUT. Here, we use an example, a D-type flip-flop, to illustrate the test pins concept. Then, in Section 5.3.2, we describe the waveform generator procedure by which we actually construct the waveform utilizing all the WAVES building blocks. Here, we demonstrate how the frame set array defines the frame sets needed for testing the UUT. We also explain the role of each building block in constructing a complete waveform. The same D Flip Flop example is expanded to explain the waveform generator procedure. Now, we begin with the test pins.

5.3.1 Test Pins

The test pins are the first device-specific WAVES building block that needs to be defined via a WAVES file. In a typical design or test environment, the stimulus test data is applied to input pins and we observe the response from the output pins of the UUT. In the WAVES-VHDL environment, we use the test pins to define input and output pins to which the waveform applies. The *test pins* are the names of pins that are used to test or exercise the UUT. They simply list all the input and output pins of the UUT. In general, the test pins names should correspond as closely as possible to conventional pin names for the UUT. We use the test pins to drive (stimulate) the device and to check the response of the device with respect to the stimulus which is applied.

Given this, and the nature of the WAVES building blocks we described in Chapter 4, we can easily relate the test pins to the other WAVES elements. Basically, a slice of a waveform can be specified as a list of pin codes, with one code per test pin, where each code refers to a frame to be applied to the corresponding pin. In order to implement this waveform concept, we need to associate each pin code with a frame, and this association is accomplished in the frame set definition, as we described in Chapter 4. We also need to associate the frame sets and the test pins to identify frames to be used for each test pin. This association is accomplished in the frame set array definition, which we will explain when we describe the waveform generator procedure in Section 5.3.2. These two levels of association (pin code to frame, frame sets to test pins) allow us to specify a segment of a waveform (in a time slice) across all the test pins as a list of pin codes.

Now, let us take a look at an example illustrating exact syntax of the test pins package declaration. As shown in Figure 5-4, the UUT (a D-type Flip-Flop) has four pins for external interface: CLK, D, Q, and Q_BAR. Subsequently, Figure 5-5 shows a **UUT_Test_Pins** package declaring the **Test_Pins** type for the D Flip-Flop.

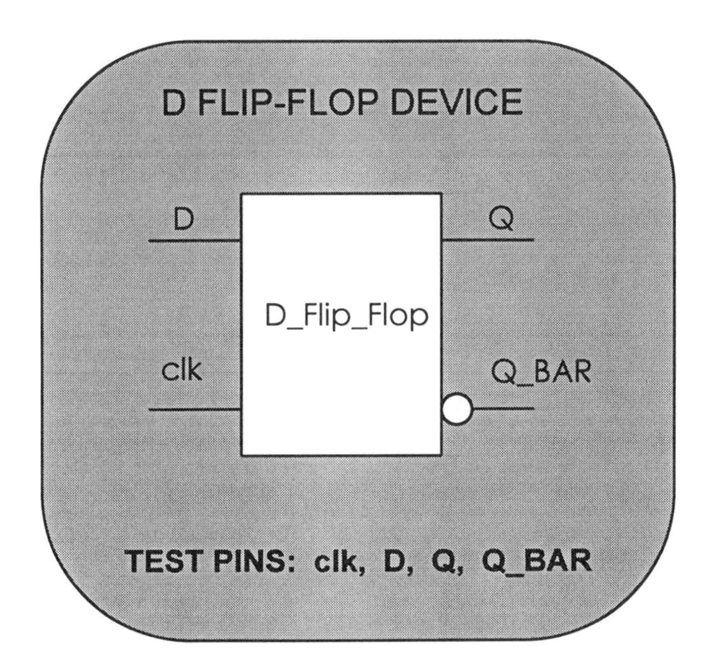

Figure 5-4. Test Pins and Pins for the D-type Flip-flop

```
PACKAGE UUT_test_pins IS

-- The test pins declaration
  TYPE test_pins IS (clk, D, Q, Q_BAR);
END UUT_test_pins;
```

Figure 5-5. Test_Pins package declaration for the D-type flip-flop

In ***summary of our test pins***, we can see that external interfaces to the UUT, such as applying stimulus and observing responses, are accomplished through the test pins package declaration, which is device-specific. Next, we discuss another WAVES building block that needs to be implemented as a WAVES file - the waveform generator procedure.

5.3.2 The Waveform Generator Procedure (WGP)

The Waveform Generator Procedure (WGP) is the second device-specific WAVES building block that we need to define within the WAVES files, to support a simulation environment. As the name implies, the WGP, which is a subprogram or collection of subprograms, constructs the waveform and supplies it to the UUT. In general, the WGP collectively reads in the external files (associating pin codes and slice starting times), interprets their contents, and constructs the waveform utilizing the WAVES building blocks we described in Chapter 4. During this waveform construction process, the WGP utilizes some functions and types defined in other packages, in addition to the WAVES building blocks. Therefore, we need to make these packages as well as the WAVES building blocks visible to the WGP. We use the ***context clauses***, which consist of ***library*** and ***use*** clauses, to ensure the visibility of the WAVES building blocks and other packages. Next, we discuss the context clauses required for the WGP, after which we will present the WGP in more detail.

The WGP requires visibility of three WAVES standard packages; **Waves_Standard**, **Waves_Interface**, and **Waves_Objects**. These packages define all the types required to create the WAVES dataset (in Level 1, we address Level 2 WAVES in later chapters). These packages also provide a controlled interface (functions and procedures) to simulation constructs to compensate for the elimination of those constructs from the WAVES syntax. The WGP also reads external files and the external file I/O requires a TEXTIO package in the STD library. The WGP utilizes the WAVES building blocks, frame sets and pin codes, defined in the **Waves_1164_Frames** and **Waves_1164_Pin_Codes** packages, respectively. In addition, the test pins we defined in Section 5.3.1 need to be associated with the frame sets. Therefore, we need to make the package **Test_Pins** visible to the WGP.

In a VHDL-WAVES environment, all packages reside in the design libraries. Consequently, when we reference the packages, we need to specify the library in which each package reside. This is accomplished by using the ***library*** and ***use*** clauses. For example, to reference the package **Waves_Standard** in the library **WAVES_STD** we need to have the context clauses "LIBRARY **WAVES_STD**;" and "USE **WAVES_STD.Waves_Standard.All**;". The required context clauses for the WGP are shown in Figure 5-6. Here, all our packages reside in four different design libraries. They are **WAVES_STD, STD, WAVES_1164, and Work**. Here, we notice that the library clauses for the STD and Work library are missing. This is because the STD library is built in VHDL and the Work library is the name defined in VHDL as a default library. These are the two exceptions that do not require the library clause. All other libraries must accompany the library clause before any packages within the libraries can be used.

```
LIBRARY WAVES_STD;
USE WAVES_STD.WAVES_Standard.all;
USE STD.textio.all;
LIBRARY WAVES_1164;
USE WAVES_1164.waves_1164_frames.all;
USE WAVES_1164.waves_1164_pin_codes.all;
USE WAVES_1164.waves_interface.all;
USE work.waves_objects.all;
USE work.uut_test_Pins.all;
```

Figure 5-6. Context Clause for the Waveform Generator Procedure

Now, we are ready to discuss the package WGP. The actual package and package body of the WGP is preceded by the context clauses which were described previously. We will use the D-type Flip-Flop example we started in Section 5.3.1 to illustrate the WGP. We begin with actual package declaration for the WGP.

The package declaration for the WGP of the D Flip-Flop is shown in Figure 5-7. The name of the package which contains the actual WGP is **WGP_D_Flip_Flop**. It declares the WGP named "***waveform***". The waveform generator procedure, waveform, uses a signal WPL, which is **Waves_Port_List** type, to communicate the waveform information with the UUT during test or simulation. The type **Waves_Port_List** is defined with respect to the test pins in the **Waves_Objects** package. In other words, the signal WPL has the same number of test ports as the test pins of the UUT. For the D Flip-Flop example, there are four test ports since the UUT contains four pins: clk, D, Q, Q_BAR. Two input test ports (clk and D) are used to stimulate the UUT and two output test ports (Q and Q_BAR) are used to provide expected results to the testbench. Now, let us look at the package body which implements the actual WGP.

```
-- Package declaration for the WGP
PACKAGE WGP_d_flip_flop is

    -- Actual Waveform Generator Procedure declaration
     PROCEDURE  waveform(SIGNAL WPL : inout WAVES_PORT_LIST);
END WGP_d_flip_flop;
```

Figure 5-7. Package declaration for the WGP

The complete *body* of the **WGP_D_Flip_Flop** package is shown in Figure 5-8, which appears later in this section. This package body contains the WGP which is the most important element among the WAVES dataset elements. As the name implies, the WGP constructs a complete waveform by putting all the WAVES building blocks together. Before presenting this entire package, we first work our way through it sequentially, and present the way the WGP constructs this waveform and the declarations required to support this process. First of all, the WGP reads test vectors (pin codes) from an external file. Hence, we need to specify the external file that the WGP is reading. This is accomplished by the following declaration.

```
-- Specify the external test pattern file
FILE vector_file : TEXT is in "vector.txt";
```

Here, we specify the external file called "vector.txt" and it is accessed with a logical file name "**Vector_File**" by the WGP. Next, we define a variable called vector which is used to store a *slice file* read from the external file. The declaration of the variable is shown below:

```
VARIABLE vector : FILE_SLICE := NEW_FILE_SLICE;
```

Next, we specify the pinsets or group the test pins that can be described by the same frame set (i.e., waveform shape). For example, we might want to observe the output signals, Q and Q_bar, at the same window time frame after these output signals reached a stable condition. Therefore, we defined a pinset constant, ***outputs,*** to group Q and Q_bar output pins. On the other hand, the input signals, D and CLK, can be better described by using different waveform shapes. Consequently, we did not group these input pins under the same pinset. The grouping of the test pins are not required, however, it is strongly recommended because the grouping enhances the readability of the WAVES codes and it simplifies the definition of the frame set array which we will discuss next. The following code demonstrates the logical grouping of the test pins for the **D Flip_Flop**:

```
-- Specify and group the test pins for testing of UUT
  CONSTANT outputs: pinset:= new_pinset((Q, Q_bar));
  CONSTANT inputs: pinset:= new_pinset((D));
```

The next declaration, required in supporting the waveform generation process, is the **Frame_Set_Array**. The frame set array definition associates the grouped and individual test pins with the applicable waveform shapes (i.e., frame sets) to be used in constructing the waveform. It also provides the edge transition(s) of each waveform shape to further define frame sets. We recall from Chapter 4, the frame sets in the library **WAVES_1164** were defined using time variables such as t1 and t2. The frame set array definition assigns actual times to these variables. We utilize the WAVES building blocks, developed in Chapter 4, for the frame set array definition. The following code shows the frame set array definition for the **D Flip_Flop**:

```
-- Define the frame set array using frame set elements
   CONSTANT vector_FSA : Frame_set_array :=
     New_frame_set_array(Pulse_high( 5 ns, 15 ns), clk) +
     New_frame_set_array(Non_return( 0 ns), inputs) +
     New_frame_set_array(window( 10 ns, 20 ns), outputs);
```

In this example, the clk signal utilizes the pulse high (PH) drive waveform format which has edge transitions at 5 ns and 15 ns with respect to the beginning of each slice. Also, the D input signal is associated with the non return (NR) format which has an edge transition at the beginning of the slice. Likewise, the pinset outputs (i.e., Q and Q_bar) are associated with the windows compare format which has edge transitions at 10ns and 20ns with respect to the beginning of each slice.

Finally, a variable "timing" is declared to organize the **Frame_Set_Array** definition into **Time_Data** format so that the WGP can utilize this information when it constructs the waveform. The following shows the declaration of the variable "timing" which is type **Time_Data**:

```
VARIABLE timing : time_data := new_time_data(vector_fsa);
```

So far, we have described the necessary declarations to associate each test pin on the UUT with a unique frame set to be used during the waveform construction process. Now, if we have a pin code for each test pin we can unambiguously define a frame (i.e., a segment of the waveform) for each test pin. Recall that frames in the frame set are indexed by the pin codes. The pin codes for each test pin are captured in the external file. Hence, we simply need to read the pin codes from the external file and construct the waveform. This process is accomplished by the following WAVES codes (i.e., the body of the WGP):

```
loop
   -- read a vector or pin codes from the external file
  READ_FILE_SLICE (vector_file, Vector);
  exit when vector.end_of_file;
  -- Construct a part of waveform
  -- and apply via WAVES port list
  apply(wpl, vector.codes.all,
        Delay(vector.fs_time), timing);
end loop;
```

In this example, the **Read_File_Slice** function reads a slice of pin codes from the external file into the variable vector which is type **File_Slice**. The variable Vector contains the pin codes for all the test pins and also the duration of the slice. Then the APPLY function utilizes this variable "vector" and the frame set array information organized in variable "timing" events in constructing a slice of waveform. This waveform is then applied to the UUT via the **Waves_Port_List**. As we mentioned previously, the **Waves_Port_List** is a signal type that is used to communicate the waveform information with the UUT for a test.

The waveform generation process we have just described continues until the end of the external file is reached. The constructed waveform is used to stimulate the UUT and to check the corresponding response of the UUT.

At this point, we have discussed all the pieces that make up the WGP, and illustrated the package body. To complete our picture of the WGP, Figure 5-8 shows the complete WAVES code for the package **WGP_D_Flip_Flop**, including context clauses, declarations, and the package body once again, now in its proper place.

```
--  Context clause to make other packages visible.
--
LIBRARY WAVES_STD;
USE WAVES_STD.WAVES_Standard.all;

USE STD.textio.all;
LIBRARY WAVES_1164;
USE WAVES_1164.WAVES_1164_frames.all;
USE WAVES_1164.WAVES_1164_pin_codes.all;
USE WAVES_1164.WAVES_interface.all;
USE work.WAVES_objects.all;
USE work.UUT_test_Pins.all;

-- Package declaration for the WGP
PACKAGE WGP_d_flip_flop is
     PROCEDURE  waveform(SIGNAL WPL : inout WAVES_PORT_LIST);
END WGP_d_flip_flop;

--------------------------------------------------------
-- Package body for the WGP
PACKAGE BODY WGP_d_flip_flop is

--  The name of the WGP
    PROCEDURE  waveform(SIGNAL WPL : inout WAVES_PORT_LIST) is

       -- Specify the external test pattern file
       FILE vector_file : TEXT is in "vector.txt";
       VARIABLE vector : FILE_SLICE := NEW_FILE_SLICE;

       -- Specify Input and output pins for testing of UUT
```

```
          CONSTANT outputs: pinset:= new_pinset((Q, Q_bar));
          CONSTANT inputs: pinset:= new_pinset((D));

        -- Define the frame set array using frame set elements
          CONSTANT vector_FSA : Frame_set_array :=
            New_frame_set_array(Pulse_high( 5 ns, 15 ns), clk) +
            New_frame_set_array(Non_return( 0 ns), inputs) +
            New_frame_set_array(window( 10 ns, 20 ns), outputs);

       VARIABLE timing : time_data := new_time_data(vector_fsa);

       BEGIN
         loop
            -- read a vector or slice from the external file
           READ_FILE_SLICE (vector_file, Vector);
           exit when vector.end_of_file;
            -- Construct a part of waveform and apply via
            -- WAVES port list
            apply(wpl, vector.codes.all,
                       Delay(vector.fs_time), timing);
         end loop;
       END waveform;
END WGP_d_flip_flop;
```

Figure 5-8. WGP of the D-type flip_flop

In ***summary of our waveform generator procedure***, we see that the WGP is the most important element in putting a waveform together, utilizing the basic WAVES building blocks. This constructed waveform is used to stimulate the UUT and to compare the actual response of the UUT with the expected response. We will present more examples of the WGP in later chapters.

In ***summary*** of the WAVES Files, we have captured some of the essential WAVES building blocks in the WAVES file. These building blocks are: (1) a logic value system considering of the Logic Values and a Value Dictionary, (2) Pin Codes, (3) Test Pins, (4) Frames and Frame Sets, and (5) a Waveform Generator Procedure (WGP). In this section, we have concentrated upon the device-specific building blocks, the test pins, and the WGP. The other, reusable, building blocks were discussed in Chapter 4 for the IEEE standard logic simulation environment. Recall that these reusable building blocks are included in the **WAVES_1164** library and they simplify the WAVES dataset generation process enormously. We will introduce the simplified WAVES dataset generation process in a later section of this chapter. Now, we present the next principal WAVES dataset element - the external files.

5.4 The External Files

In the previous sections of this chapter, and in Chapter 4, we have described the WAVES files which are used to construct WAVES building blocks. In this section, we describe our *second principal WAVES dataset element* - the External Files. An external file contains information essential to build a complete waveform. More precisely, in WAVES Level 1, the external file contains pin codes and slice timing information. Each line in an external file represents a time slice in a waveform across all signals. As we discussed in Section 5.3.2, the waveform generator procedure (WGP) reads pin codes and the slice time information one line at a time from the external file, and constructs a corresponding slice of the waveform utilizing the building blocks. Then, this slice of waveform is passed to the testbench via the **Waves_Port_List** for WAVES-VHDL simulation.

Actually, there is more to this external file than we describe here. So much more, in fact, that we have devoted an entire chapter to it (Chapter 7). Here, however, we present only a very brief explanation of the external file, with an example, which is sufficient to provide a basic understanding of its structure and its role in generating a waveform. A more extensive and complete explanation of the external file will be deferred until Chapter 7. Now, we begin with an external file for the same D-type Flip-Flop example we began earlier, which we illustrate in Figure 5-9. The name of the file is "vector.txt" which is consistent with the external file declaration of the WGP we described in Section 5.3.2.

```
%clk  D   Q   Q_BAR          <-- (comment)
  1   1   1    0 : 20 ns;   <--  (Slice 1)
  0   0   1    0 : 20 ns;   <--  (Slice 2)
  1   0   0    1 : 20 ns;   <--  (Slice 3)
  1   1   1    0 : 20 ns;   <--  (Slice 4)
  1   0   0    1 : 20 ns;   <--  (Slice 5)
                ↑
           This space must exist
```

Figure 5-9. Test vectors for D flip-flop

Now let us examine the particulars of the external file in this figure. First, for this example, we begin with a comment line which is denoted with a "%" symbol at the beginning of the line. We used the comment line, in this case, to explain that each column of pin codes corresponds to each pin in the test pins. More generally, comments can be used to clarify and document the nature of the test and test vectors (pin codes). We can add as much commenting we deem necessary, as long as each line begins with the "%" symbol. (A comment line must start with "%" symbol). Next, this external file contains five test vectors for the D-type flip-flop. Here, the order of the pin codes in each slice is very important to build the waveform properly. The order of the pin codes in each slice must match with the order of test pins. In the declarations we discussed earlier, and illustrated in Figure 5-5, we declared the test pins in the following order: clk, D, Q, Q_BAR. Therefore, we must ensure that the first column of the pin codes represents test vectors for the *clk* signal. Similarly, the second column is for *D*, the third column is for *Q*, and the fourth column is for *Q_bar*. The length of each slice is 20 ns. Syntactically, each set of pin codes and the length of the slice must be separated by a colon ":" and a space " ". Also, each slice must end with a semicolon ";".

Now, we have all the necessary pieces to construct a complete waveform. We can see how these pin codes work with the frame sets to perform this construction, using the *clk* signal to explain the construction process. First, referring to Figure 5-10, we get a pin code "1" from the external file for slice 1. Returning to the frame set array definition in the WGP, previously illustrated in Figure 5-9, we note that the *clk* pin is associated with **Pulse_High** frame set or waveform format. This frame set array definition, shown below, also defines the edge transition event times for the frame set, **Pulse_High**. We note that these edge transitions occurs at 5ns and 15ns with respect to the beginning of the slice.

```
CONSTANT vector_FSA : Frame_set_array :=
          New_frame_set_array(Pulse_high( 5 ns, 15 ns), clk) +
          New_frame_set_array(Non_return( 0 ns), inputs) +
          New_frame_set_array(window( 10 ns, 20 ns), outputs);
```

Recall, from Chapter 4, that the frame set contains a unique frame for each pin code. In Section 4.3.3.2 we described the **Pulse_High** frame set which, when included in a Frame Set Array, described edge transition event times relative to the slice starting times. Now, we have a pin code and the frame set with absolute edge transition times. Hence, we can unambiguously define a segment of the waveform for the *clk* pin as shown in Figure 5-10.

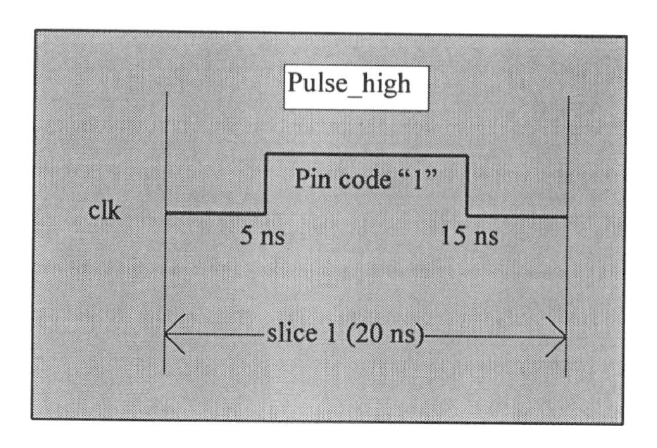

Figure 5-10. Segment of the waveform for clk pin in slice 1.

Likewise, if we repeat this process for the entire set of pin codes in the external file, a complete waveform can be constructed as shown in Figure 5-11. Fortunately, we do not have to actually exercise this process because the VHDL simulator can build the waveform quickly and efficiently. We simply need to specify the waveform utilizing the WAVES building blocks.

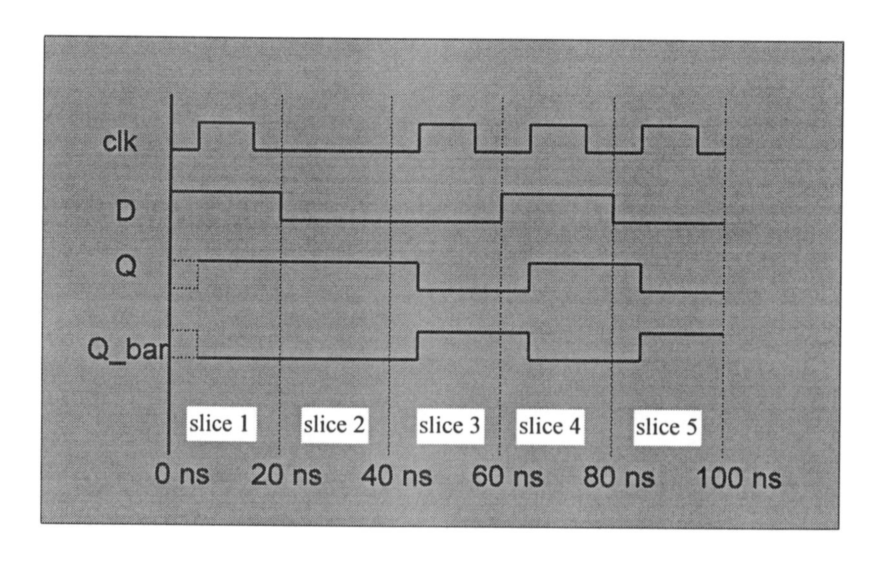

Figure 5-11. Complete Waveform for the D-type flip-flop

In ***summary of our external file,*** we have introduced a mechanism to capture the waveform information into an external file which, when combined with the frame set array definition, unambiguously describes all slices of a waveform. Here, we must realize that the pin codes in the external file alone cannot generate the waveform. The pin codes must interact with other WAVES building blocks such as frame sets and test pins. These interactions are managed by the waveform generator procedure we discussed in Section 5.3.2. We will present more examples of the external file, including some interesting variations, in later chapters. Now, we present the last principal WAVES dataset element - the Header File.

5.5 The Header File

So far we have described two principal WAVES elements; the WAVES files and external file. The WAVES files we described in Section 5.3 are used to construct WAVES building blocks, and the external file we outlined in Section 5.4 is used to provide test vectors (pin codes) for indexing the building blocks. As we described in previous sections, these two principal elements work together to construct a complete waveform. In fact, information provided by these two elements were sufficient to create a waveform. Then, we might wonder why the third element, the header file, is needed. What does it do? We answer these questions in this section.

We use our *third dataset element*, the WAVES dataset header file, to specify information that is not available within the WAVES files or the external files, yet that is necessary or useful to describe the WAVES dataset completely. The header file serves three distinct functions: (1) identifies the WAVES dataset, (2) describes how the dataset is to be assembled, and (3) identifies the external file which contains the pin codes (vectors) to be used as part of the dataset. In addition, the header file specifies the names of the waveform generator procedures, which utilize one or more of the waveform generation operations discussed in Section 5.3, and provides any additional comments deemed necessary.

This documentation and identification information contained in the header file is very important and it is an integral part of the WAVES dataset. If we recall the original purpose of WAVES, the importance of the header file becomes obvious. As we mentioned in Chapter 2, WAVES was developed as an exchange specification for waveforms and test vectors among various organizations. To facilitate and enhance communications between organizations, it is essential to have the documentation and identification information of the WAVES dataset. This documentation and identification information also can be very useful if we have to revisit the design project we completed several months ago. For example, we may not remember the exact compilation order of the WAVES dataset. The good news is that we can get this information from the header file. Now, we realize the importance of the header file. We should also realize that the header file does not need to be analyzed for the WAVES-VHDL simulation because it contains only useful documentation and identification information, not design or test information.

For a more formal and syntactical description of the header file, we should consult Chapter 2 of IEEE STD 1029-1991. Now, let us describe the header file which consists of four sections: the Dataset Identification, the WAVES File Identification, the External File Identification, and the Waveform Identification.

5.5.1 Dataset Identification

The Dataset Identification section supports our first header file function, to identify the WAVES dataset. Using this section, we may provide useful information about the name of the dataset, the author, the release date, the origin, and a device identifier. We use the following fields to identify the dataset:

Title	A descriptive name for the data
Device_id	The Unit Under Test (UUT) to which the dataset is to be applied
Date	A timestamp
Origin	The originating Computer Aided Design (CAD) or Automatic Test Equipment (ATE) environment of the dataset
Author	The person(s) or organization(s) producing the dataset
Other	Comments relating to the data.

The following is an example of the dataset identification. In this example, assume that the original dataset was created by a "Company X Design Team" and that it was later modified by a "Company Y Design Team."

Data Set Identification Information from Company X:

```
TITLE          Test Vector for  D Flip-Flip with Preset
               and Reset
DEVICE_ID      D_flip_flop
DATE           Wed Sep  6 15:09:10 1995
ORIGIN         Company X Design Team
AUTHOR         (Company or Person)
AUTHOR         (May be Multiple - Companies or People)
```

Revision to the original dataset, by Company Y:

```
DATE        Wed Oct  11 16:09:10 1995
ORIGIN      Modified by Company Y Design Team
AUTHOR      (Who did it, Company or Person)
OTHER       (Any general comments which are useful)
OTHER       Built Using the WAVES-VHDL 1164 STD
            Libraries
```

Hence, we see that the dataset identification section of the WAVES header file provides the means for attaching useful information, which may be altered as the dataset evolves from user to user, or from organization to organization.

5.5.2 WAVES File Identification

The WAVES File Identification section supports our second header file function, to identify and describe how the dataset is to be assembled (the compilation order). We use this section to provide the name of the WAVES source files and standard packages in the correct order for analysis. To accomplish this, we use several keywords provided in the WAVES LRM (IEEE 1029-1991). The keywords are **WAVES_FILENAME**, **WAVES_UNIT**, **WAVES_UNIT_SPEC**, and **WAVES_UNIT_BODY**. Now let us discuss how these keywords are used to identify the WAVES dataset.

The keyword "**WAVES_FILENAME**" is used to identify the name of the WAVES source files. It also provides a mechanism for specifying the library into which each WAVES file is analyzed. Syntactically, the name of the file, which depends on the local operating system's file name convention, must be proceeded by a keyword "**WAVES_FILENAME**" and followed by the library name into which the WAVES file is analyzed. The following shows an example of a WAVES file identification. Here, and in subsequent examples, the explanation of each field is indicated in *italics.*

`WAVES_FILENAME`	`/wav_pins.vhd`	`WORK`
(keyword)	*(file name)*	*(library name)*

Now we can describe the use of the keywords: "**WAVES_UNIT**", "**WAVES_UNIT_SPEC**", and "**WAVES_UNIT_BODY**". A WAVES file may contain a reference to an already-existing WAVES file in the design library of the organization receiving the dataset. These keywords are used to reference the pre-existing WAVES units and to identify which parts of the package shall be analyzed.

WAVES_UNIT	Both the package declaration and the package body to be analyzed

WAVES_UNIT_ SPEC	Only the package declaration to be analyzed
WAVES_UNIT_BODY	Only the package body to be analyzed

The following example shows the use of the keyword **WAVES_UNIT**, which references both the package declaration and the package body. This example implies that the package **Waves_Objects** resides in the library WORK and a WAVES file which references the package **Waves_Objects** will be analyzed into the WORK library. Again, the explanation of each field is indicated in *italics.*

WAVES_UNIT	WAVES_OBJECTS	WORK
(keyword)	*(package name)*	*(library name)*

We may also use this section to specify the packages in a soft standard library which make up a part of the WAVES dataset. In general, these packages contain reusable WAVES elements and are referenced by other WAVES building blocks. This is accomplished by using the context clauses (such as context_item). The context clauses utilize the keywords "library" and "use" to specify the library and the relevant packages. The following example illustrates the use of these context clauses. In this example, the context clauses specify the library **WAVES_1164** which contains the packages **Waves_1164_Pin_Codes**, **Waves_1164_Logic_Value**, **Waves_Interface**. In general, this **Context_Item** is textually inserted at the beginning of the WAVES units, which reference the packages above, prior to analysis of that unit. We have discussed the library **WAVES_1164** in Chapter 4, and its use earlier in this chapter. Therefore, this context clause should be familiar by now.

```
library  <--(keyword)       WAVES_1164;
use <--(keyword)
WAVES_1164.WAVES_1164_Pin_Codes.all;
use <--(keyword)
WAVES_1164.WAVES_1164_Logic_Value.all;
use <--(keyword)
WAVES_1164.WAVES_Interface.all;
```

Now, having discussed all the individual elements of the file identification section in a header file, we present a complete example of a file identification section. We should also realize that the following code describes the order of analysis for the WAVES units.

WAVES file identification (sometimes called Dataset Construction Information):

```
WAVES_FILENAME    /wav_pins.vhd                  WORK
```

```
library            WAVES_1164;
use                WAVES_1164.WAVES_1164_Pin_Codes.all;
use                WAVES_1164.WAVES_1164_Logic_Value.all;
use                WAVES_1164.WAVES_Interface.all;
use                WORK.DUT_Test_pins.all;
WAVES_UNIT         WAVES_OBJECTS                      WORK
WAVES_FILENAME     ./wav_gen.vhd                      WORK
```

In summary, we see that the WAVES File identification section provides the means for properly ordering the WAVES source files and standard packages for analysis.

5.5.3 External File Identification

The External File Identification section supports our third header file function, to identify the vectors to be used as part of the dataset. Here, we may specify a logical name for an external file which is accessed from a WAVES dataset. The format of the external file identification is that a physical file name is preceded by a keyword "**EXTERNAL_FILENAME**" and followed by a logical file name. Zero or more external file statements may be in a WAVES dataset header file. The following is an example of such an external file identification:

```
EXTERNAL_FILENAME  vectors.txt                 VECTOR_file
```

As such, we can always be assured of an explicitly-defined external file for WAVES dataset accessing.

5.5.4 Waveform Identification

The Waveform Identification section of the header file supports our need to specify the names of the Waveform Generator Procedures (WGPs). We use this section to identify a waveform generator procedure, and the library and package in which the procedure is declared. The procedure name is composed of the library name, the package name, and the procedure name, which are all separated by dots ("."), and preceded by a keyword "**WAVEFORM_GENERATOR _PROCEDURE**". One or more waveform identification statements may be in a WAVES dataset header file. The following is an example of this waveform identification:

```
WAVEFORM_GENERATOR_PROCEDURE WORK.WAVES_d_flip_flop.waveform
```

Hence, our waveform identification section provides the means to explicitly associate the library, the package, and the waveform generator procedure.

Now, having discussed all the elements of a header file, we present a complete example of a header file for the same example, our D-type flip-flop:

```
-- *****************************************************
--
-- ******** Header File for Entity: d_flip_flop
--
-- *****************************************************
-- *****************************************************
--
-- Data Set Identification Information
--
TITLE      Test Vector for  D Flip-Flop with Preset and
           Reset
DEVICE_ID  D_flip_flop
DATE       Wed Sep  6 15:09:10 1995
ORIGIN     Company X Design Team
AUTHOR     Company or Person
AUTHOR     Maybe Multiple ... Companies or People

--  Revision to original dataset
DATE         Wed Oct  11 16:09:10 1995
ORIGIN       Modified by Company Y Design Team
AUTHOR       Who did it, Company or Person
OTHER        Any general comments useful
OTHER        Built Using the WAVES-VHDL 1164 STD  Libraries

-- Dataset Construction Information
--
WAVES_FILENAME   ./wav_pins.vhd                     WORK
library          WAVES_1164;
use              WAVES_1164.WAVES_1164_Pin_Codes.all;
use              WAVES_1164.WAVES_1164_Logic_Value.all;
use              WAVES_1164.WAVES_Interface.all;
use                    WORK.DUT_Test_pins.all;
WAVES_UNIT             WAVES_OBJECTS                WORK
WAVES_FILENAME         ./wav_gen.vhd                WORK

-- External file identification
EXTERNAL_FILENAME      vectors.txt          VECTOR_file

-- Waveform Identification
WAVEFORM_GENERATOR_PROCEDURE WORK.WAVES_d_flip_flop.waveform
```

In ***summary of the WAVES header file dataset element***, we see that it includes four principal sections, for the purpose of performing four functions: to identify the WAVES dataset, to describe how the dataset is to be assembled, to identify the test vectors to be used as part of the dataset, and to identify the waveform generator procedure. The information contained in the header file is very essential in maintaining the WAVES dataset and it facilitates the communication between organizations. Next, we will introduce the WAVES dataset development procedure which summarizes the relationships among the WAVES building blocks described so far, and all the supporting WAVES standard packages.

5.6 The WAVES Dataset Development Procedure

As we illustrated at the beginning of this chapter, in Figure 5-1, the WAVES dataset is typically made up of many packages or files. In addition, the WAVES dataset requires several standard packages. We have already introduced all the WAVES building blocks and the standard packages in Chapters 4 and 5. In this section, we provide *cookbook-style* instructions which summarize the WAVES dataset generation. In this process, we concentrate upon the analysis order of the WAVES building blocks and the standard packages. Since VHDL/WAVES is a strongly-typed language, all standard packages and Waves files (i.e., building blocks) must be analyzed in a certain order to ensure the visibility requirement imposed by the language. Here, we will utilize the libraries **WAVES_STD** and **WAVES_1164** to simplify the dataset generation process. Figure 5-12 illustrates the WAVES dataset development procedure which consists of seven steps. Here, the first two steps, which establish the libraries **WAVES_STD** and **WAVES_1164**, need to be done only once for a given simulation or testing environment. The next five steps accommodate the device-specific information. Hence they need to be repeated or customized for each different UUT. In this sense, the WAVES dataset generation is simple and it requires only five steps for each UUT. Now, we begin with the first step, to establish the library **WAVES_STD**.

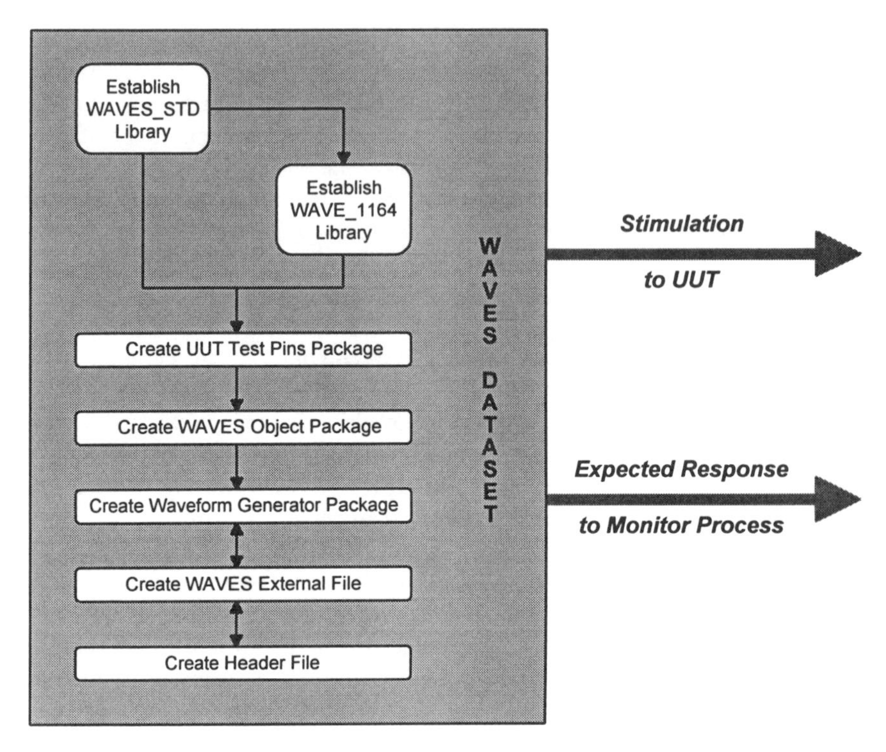

Figure 5-12. WAVES Dataset Development with WAVES_1164 Library

5.6.1 The WAVES_STD Library

The first step of the WAVES dataset development procedure is to establish the library **WAVES_STD**. This library contains two packages; **Waves_Systems** and **Waves_Standard**. The **Waves_Systems** package defines the basic types that are used by the three WAVES standard packages (**Waves_Standard**, **Waves_Interface**, and **Waves_Objects**) as system-dependent information. In general, we do not need to worry about the contents of this package because this package contains only system-dependent information that supports the WAVES standard package. The **Waves_System** package developed by the **Test Analysis and Standardization Group (TASG)** of the IEEE is included on the companion CD-ROM. The name of the file is "***Wav_Std.Vhd***" located in the ***WAV_PACK*** directory. Since this package has been proven to work, its use is strongly recommended.

The second package in the **WAVES_STD** library is the **Waves_Standard**. The **Waves_Standard** package defines the various types and functions that are useful in defining the WAVES building blocks, which we described in Chapter 4. The formal description of the package is presented in Section 7.1 of IEEE STD 1029.1-1991. However, implementation of the package (package body) is not defined. We again provide a complete **Waves_Standard** package, including implementation, which was developed and evaluated by TASG of the IEEE, on the companion CD-ROM. This package is appended at the end of the **Waves_Systems** package in the same file "***Wav_Std.Vhd***" located in the ***WAV_PACK*** directory. Again, since this package has been proven to work, we are saved the trouble of developing our own.

At this point, we can use a VHDL analyzer to compile the file **Wav_Std.Vhd (Waves_System** and **Waves_Standard** packages) into the **WAVES_STD** library. As we mentioned above, the library **WAVES_STD** needs to be established only once for a given simulation environment.

5.6.2 The WAVES_1164 Library

The second step of the WAVES dataset development procedure is to establish the library **WAVES_1164**. We have already described this library in Chapter 4, where we indicated that we needed a package called **Waves_Interface**, whose explanation was deferred to a later chapter. This is now the place to examine this **Waves_Interface** package.

The **Waves_Interface** package provides essential types and functions to help define some of the WAVES building blocks and to facilitate the interface among these building blocks. For example, this package defines the types **Event_Time_Basis**, **Delay_Time_Basis**, **Frame**, **Frame_List**, **Frame_Set**, **Logic_Map**, and **Frame_Set_Array** by declaring subtypes of the private types declared in the package **Waves_System**, which we described in Section 5.6.1. The **Waves_Interface** also declares functions used to support the construction of various frame sets and APPLY functions for a waveform. The formal description of the **Waves_Interface** package is found in Section 7.2 of IEEE STD 1029.1-1991. However, implementation of the package (package body) is not provided. Therefore, we include a complete **Waves_Interface** package, including implementation, which was evaluated by TASG of the IEEE, on the companion CD-ROM. This package is included in a file called "**Wav_1164.Vhd**" located in the **WAV_PACK** directory. This file also contains all the reusable WAVES building blocks we discussed in Chapter 4, in proper analysis order.

At this point, we can use a VHDL analyzer to compile the file **Wav_1164.Vhd** into the **WAVES_1164** library. Again, the **WAVES_1164** library needs to be established only once for a given simulation environment because it contains reusable WAVES building blocks.

5.6.3 The UUT Test Pins Package

The next step in building a waveform is to define test pins that name the pins to which the waveform applies. We described the test pins package in Section 5.3.1. This package is device specific; hence, it must be compiled into our default work library. This package needs to be analyzed prior to analysis of the **Waves_Object** package, which requires visibility of the test pins package.

5.6.4 The WAVES Object Package

The next step in the dataset development procedure is to add necessary context clauses and compile it into our default work library. The **Waves_Object** package is one of the three WAVES standard packages. Unlike the other two standard packages, **Waves_Standard** and **Waves_Interface**, this package is dependent on the device-specific packages. In particular, it requires visibility of the test pins package which is device specific. It also requires visibility of **Waves_System**, **Waves_Interface**, **Waves_1164_Logic_Value**, and **Waves_1164_Pin_Codes** Package. Therefore, we must add the context clauses, as shown in Figure 5-13, at the beginning of the **Waves_Object** package. Since the **Waves_Object** package is dependent on a device-specific package, this package needs to be compiled into the default work library for every UUT.

```
--
use STD.TEXTIO.all;
-- Object package use TEXTIO
library WAVES_STD;
use    WAVES_STD.WAVES_SYSTEM;
Library WAVES_1164;
use WAVES_1164.WAVES_Interface.all
use WAVES_1164.WAVES_1164_Logic_Value.all;
use WAVES_1164.WAVES_1164_Pin_Codes.all;
use Work.UUT_Test_pins.all;
--
```

Figure 5-13. Context Clause of WAVES_Object Package

The **Waves_Object** package defines the private types **Waves_Port_List** and **Waves_Match_List** by declaring subtypes of the private types declared in the package **Waves_System**, which we described in Section 5.2.1. The object **Waves_Port_List** is used to communicate waveform information from waveform generator procedures to the external environment. The object **Waves_Match_List** is used to communicate waveform response information from the external environment to waveform generator procedures. In addition, this **Waves_Object** package contains functions that construct a frame set array, and communicate waveform data such as ***APPLY, TAG, and MATCH***, which we shall encounter in later chapters with the external world. It also contains definitions and functions that support the fixed-file format of the Level 1 WAVES external file interface. This package is defined in Section 7.3 of IEEE STD 1029.1-1991; however, implementation of the package (package body) is not defined in the standard. We again provide a complete **Waves_Object** package including implementation, which was developed by TASG of the IEEE, on the companion CD-ROM (file name **Wav_Obj.Vhd** located in the **WAV_PACK** directory).

5.6.5 The Waveform Generator Package

The next step in building a waveform is to define one or more waveform generator procedure(s) (WGPs). As we described in Section 5.3.2, the waveform generator procedure plays an important role in constructing a waveform. The WGP is also device-specific, therefore, it needs to be compiled into our default work library.

5.6.6 The WAVES External File

Providing the external file, which we outlined in Section 5.4, is *the sixth step of the dataset development procedure*. This file is a pure data/text file which does not need to be analyzed and it is only referenced by the waveform generator procedure, as we discussed in Section 5.3.3. Hence, it can be created anytime prior to simulation of the WGP, which means that changing the external file does not affect the compilation of other WAVES files. This feature allows us to run various tests or simulations by simply changing the content of the external file without analyzing any WAVES and VHDL files. We will provide a more extensive description of the external file in Chapter 7.

5.6.7 The WAVES Header File

Providing the header file (see Section 5.5), *the last step in building a waveform*, provides useful comments and information to document the WAVES dataset. As described in Section 5.5, the information in the header file plays an important role in facilitating communication between organizations. It also can be very useful in maintaining the WAVES dataset. The header file does not need to be compiled. We must, though, keep this file in the directory in which most of the WAVES dataset exists.

In ***summary of our dataset development procedure***, we see that it brings all the necessary elements together for the dataset. These include libraries, units for compilation, and reference files. A useful tool for assembling the dataset is included in the tools directory on the companion CD-ROM.

In ***summary of our chapter on the WAVES dataset***, we have illustrated the role of the WAVES dataset in testing/simulation environments. We described the construction process in detail, including an explanation of the WAVES dataset structure and its three principal elements. Next, we described the function of each of these three WAVES dataset elements in constructing a waveform to provide an explicit illustration of how we go about this process. Finally, we summarized this chapter with the WAVES dataset development procedure. Next, in Chapter 6, we will revisit, through an overview, the design and test environment to illustrate the role of the testbench and its relationship with the WAVES dataset. Then, we will introduce additional utility packages which facilitate the WAVES-VHDL testbench generation. Here, we will construct a complete testbench which integrates the WAVES dataset.

CHAPTER 6. COMPLETE WAVES/VHDL INTEGRATION

Configuring and executing the complete simulation

In this chapter, which concludes our basic tutorial sequence, we describe how to put everything together, to simulate the WAVES and VHDL combination and obtain the results we need to support our design and testing endeavors. We begin with a conceptual view of the functionality of the integrated WAVES-VHDL simulation system. Here, we explain how all the elements in a WAVES-VHDL testbench interface during simulation. Next, we present a group of interface functions necessary to implement the conceptual model of the testbench in the WAVES-VHDL simulation environment. The purpose and role of each function, including actual VHDL codes, are presented here, and all functions are collected in the **Waves_1164_Utilities** package. Then, we describe an implementation of a WAVES-VHDL simulation system, which utilizes the libraries **WAVES_STD** and **WAVES_1164**, with an example. Here, we demonstrate that the WAVES files necessary to test and simulate a design can be generated with minimal effort. Finally, we conclude this chapter with an additional example to further assist us in developing the WAVES dataset.

6.1 The Integrated WAVES-VHDL Simulation System

We begin with a conceptional view of the functionality of the integrated WAVES-VHDL simulation system, as we depict in Figure 6.1. As we described in Chapter 5, a WAVES dataset is typically made up of many packages or files. We can refer back to Chapter 5 if there are any questions regarding the WAVES dataset. During the simulation cycle, the WAVES dataset, which is a collection of waves building blocks and the standard packages, generates a waveform to stimulate the UUT and to provide the expected response of the UUT to the WAVES-VHDL testbench. More precisely, the testbench invokes a waveform generator which is a part of the WAVES dataset and instantiates the UUT (VHDL Model), to execute the test. The testbench also contains monitor processes to compare the expected responses to the actual responses of the UUT.

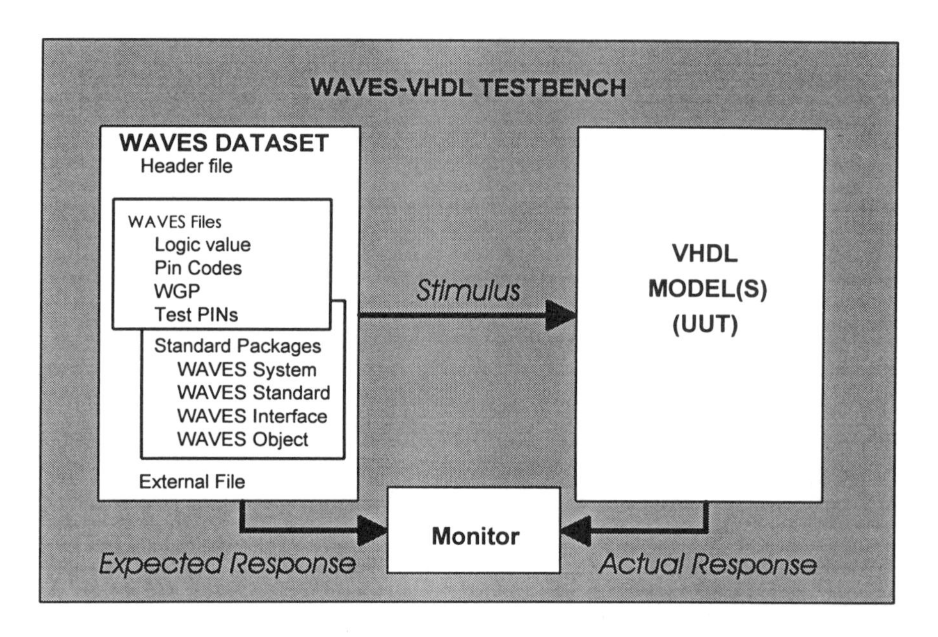

Figure 6-1. Integrated WAVES-VHDL Simulation System

In order to actually implement this conceptual model in the WAVES-VHDL simulation environment, we need to develop some interface functions. The interface functions are necessary because each element in the testbench expects different types of data. For instance, the waveform generator sends the stimulus and the expected response of the UUT in integer values during simulation, while the UUT and Monitor processes may expect the **Standard_Logic_1164** type. In the next section, we present the testbench utilities package which contains all the interface functions necessary to facilitate the WAVES-VHDL simulation.

6.2 The WAVES_1164_Utilities Package

In this section we describe the VHDL package called "**Waves_1164_Utilities** Package," which was developed to support VHDL model simulation and verification utilizing a WAVES dataset. This package builds upon the **Waves_1164_Logic_Value** package we defined in Chapter 4, and provides us with a seamless interface when utilizing VHDL and WAVES together. This set of library functions was developed to provide for the integration of WAVES into a VHDL simulation environment. The library functions provide for two different testbench aspects: first, to connect the WAVES port list signals to the model and second, to evaluate the model response for compatibility with the expected response. The premise of the functions provided in this package is that all the detailed work in developing the testbench is minimized. The functions support a simple testbench with uni-directional pins, as well as a complex bi-directional design entity. The purpose of the functions is to provide a methodology which minimizes the work required by an individual to develop testbenches, through the use of a common user library and a building-block approach. We are not restricted by this methodology, which is meant to eliminate repetitive work and reduce the learning cycle for WAVES.

Figure 6-2 illustrates the simplified view of the functionality of the testbench methodology using these packages. The stimulus waveforms which are produced by the WAVES dataset are applied to the VHDL model. The model then produces its output response based on the stimulus which was presented. The testbench contains a set of monitoring processes which verify the expected and actual response values by ensuring that they conform to the timing as specified in the WAVES dataset.

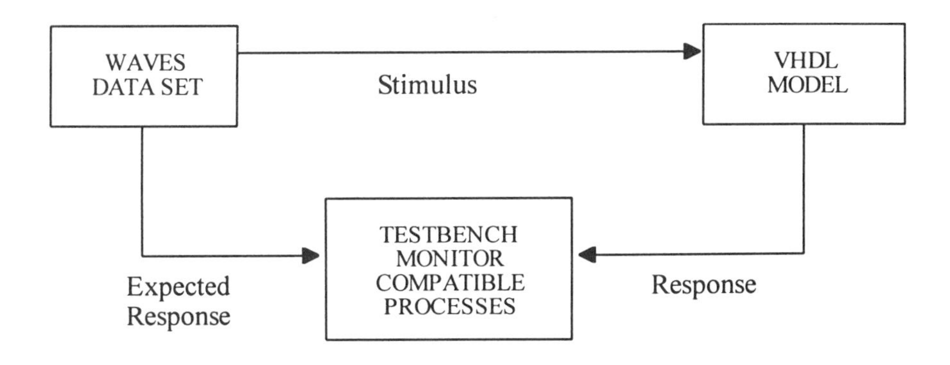

Figure 6-2. Simplified View of the Testbench

The **Waves_1164_Utilities** package contains the **Stim_1164, Expect_1164**, and **Bi_Dir_1164** functions to provide translation between the internal WAVES logic values (**Waves_1164_Logic_Value** package), as described in Chapter 4, and the IEEE STD 1164-1993 values in the users' model (the UUT). This package also contains an overloaded **compatible** function which allows us to check or verify if the predicted or expected response is "compatible" with the actual results generated by the model (see Section 6.2.2 for further discussion of the compatible function). In order to use this package, all of the VHDL design units must be compliant with the IEEE standard logic package. In Sections 6.2.1 and 6.2.2 we describe all of the functions provided in the testbench utility package. The package declarations that define the functions are shown below in Figure 6-3, and the complete implementation of the **Waves_1164_Utilities** package is included in Appendix D. This package is also included in the **Wav_1164.Vhd** file located in the **WAV_PACK** directory on the companion CD-ROM.

```
PACKAGE waves_1164_utilities IS
  ------------------------------------------
  -- Procedure and Function Declarations --
  ------------------------------------------
  --
FUNCTION stim_1164( port_element : system_waves_port)
       RETURN STD_LOGIC;
FUNCTION stim_1164( port_list : system_waves_port_list)
       RETURN STD_ULOGIC_VECTOR;
FUNCTION stim_1164( port_list : system_waves_port_list)
FUNCTION expect_1164( port_element : system_waves_port)
       RETURN STD_ULOGIC;
FUNCTION expect_1164( port_list : system_waves_port_list)
       RETURN STD_ULOGIC_VECTOR;
FUNCTION bi_dir_1164( port_element : system_waves_port)
       RETURN STD_LOGIC;
FUNCTION bi_dir_1164( port_list : system_waves_port_list)
       RETURN STD_LOGIC_VECTOR;
FUNCTION compatible( actual: STD_LOGIC;
                     expected : STD_ULOGIC )
       RETURN BOOLEAN;
FUNCTION compatible( actual: STD_ULOGIC_VECTOR;
                     expected : STD_ULOGIC_VECTOR)
       RETURN BOOLEAN;
FUNCTION compatible( actual: STD_LOGIC_VECTOR;
                     expected : STD_ULOGIC_VECTOR)
       RETURN BOOLEAN;
END waves_1164_utilities;
```

Figure 6-3. WAVES_1164_UTILITIES Declarations

Through this package description, we have completed our description of how we construct the library **WAVES_1164**, having described its other elements in previous chapters. As we mentioned very early in this text, this library's purpose is to provide a collection of reusable functions to simplify the WAVES dataset creation process, as we illustrate in Figure 6-4. In the next two sections, we describe the anatomy of the WAVES translation and compatible functions.

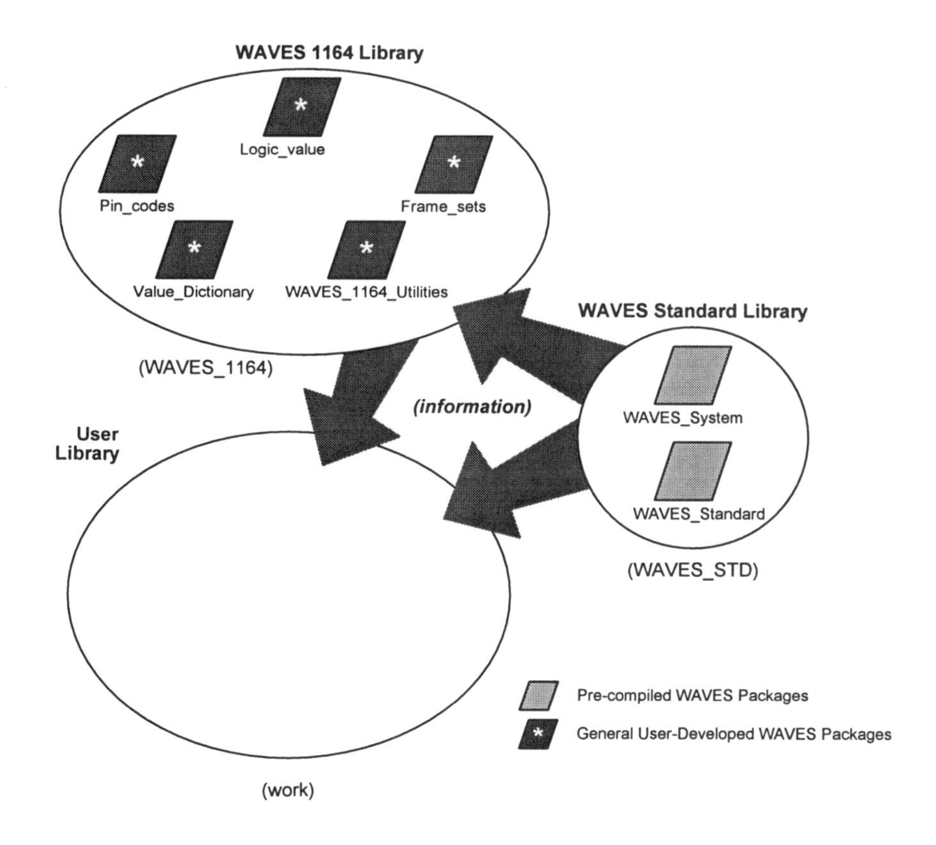

Figure 6-4. Application of the WAVES 1164 Library

6.2.1 WAVES Translation Functions

When used in simulation, the WAVES dataset generates waveforms based on the WAVES logic values we described in Chapter 4. These generated WAVES port values are not compatible with the IEEE STD 1164-1993 logic values. Instead, the WAVES port values generated during simulation are integer values. Therefore, translations are required between the WAVES port values and the UUT model's logic values.

The WAVES integer values correspond to the position of the logic value enumeration as we defined in Chapter 4. The mapping between the enumerated logic and the integer values generated on the WAVES port is shown in columns 1 and 2 of Table 6-1. The three overloaded functions **Stim_1164, Bi_Dir_1164,** and **Expect_1164** provide for translation of the WAVES port values into the IEEE STD

1164-1993 logic values to be used in a model. Table 6-1 shows the mapping values and the results that will be generated by the associated translation function.

Table 6-1. STIM_1164 Mappings

WAVES LOGIC	Waves Port Value	stim_1164 Value	expect_1164 Value	**bi_dir_1164** Value
DONT_CARE	0	'-'	'-'	'Z'
SENSE_X	1	'X'	'X'	'Z'
SENSE_0	2	'0'	'0'	'Z'
SENSE_1	3	'1'	'1'	'Z'
SENSE_Z	4	'Z'	'Z'	'Z'
SENSE_W	5	'W'	'W'	'Z'
SENSE_L	6	'L'	'L'	'Z'
SENSE_H	7	'H'	'H'	'Z'
DRIVE_X	8	'X'	'-'	'X'
DRIVE_0	9	'0'	'-'	'0'
DRIVE_1	10	'1'	'-'	'1'
DRIVE_Z	11	'Z'	'-'	'Z'
DRIVE_W	12	'W'	'-'	'W'
DRIVE_L	13	'L'	'-'	'L'
DRIVE_H	14	'H'	'-'	'H'

Figure 6-5 illustrates a testbench that would be developed for design analysis. The testbench is capable of supporting the analysis of any complex bit-level design utilizing STD logic uni-directional or bi-directional pins. The figure contains labeled blocks that identify where the functions **Stim_1164, Bi_Dir_1164,** and **Expect_1164** would be utilized in supporting the necessary logic value translations within the testbench. To further clarify how these functions are used in the construction of the testbench, we developed some examples. They are presented in Section 6.4, and in Chapter 8. We now describe these three functions individually.

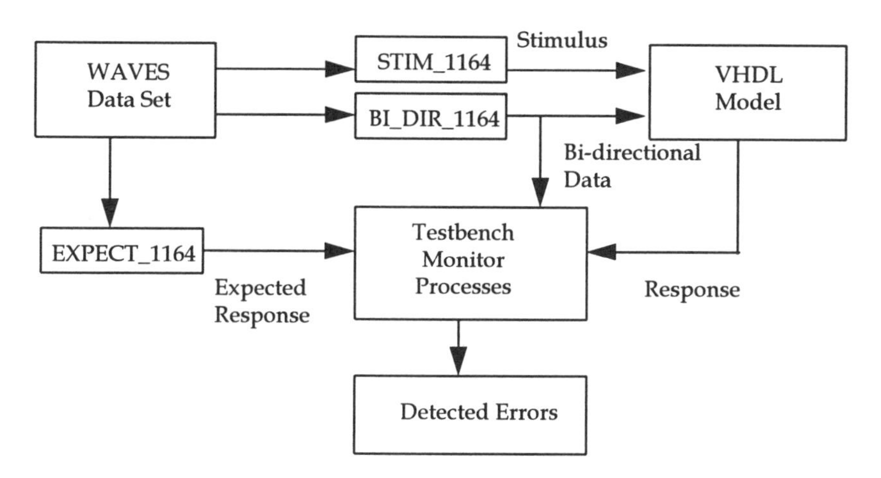

Figure 6-5. Bi-Directional Mappings

The **Stim_1164** function converts a WAVES port integer into an 1164 **Std_Logic** value. The translation function is overloaded to translate a single WAVES port integer into a **Std_Logic** bit or a list of WAVES port integers into a **Std_Logic_Vector**. The function supports both resolved and unresolved IEEE STD 1164-1993 vector types. Columns 1 and 3 of Table 6-1 shows the mapping relationships between the WAVES logic values, as we described in Section 4.3.1, and the standard logic simulation code that is produced from the WAVES port integer value. This function, as shown in Figure 6-5, is used for all model signals that are declared as input only. This function translates the WAVES port integer values into stimuli for input into the model.

The **Bi_Dir_1164** function also converts a WAVES port integer into an 1164 **Std_Logic** value. This function was developed to support the translation of inputs for bi-directional pins. It translates all of the drive values exactly the same as **Stim_1164**, when the model's port is in the input mode. In order to prevent any conflict when the model's port is in the out state, this function strips off the sense values and replaces them with 'Z', a high impedance. This translation of any sense values to high impedance allows the IEEE STD 1164-1993 bus resolution function to resolve the state of the bi-directional signal generated by the model. The translation function is also overloaded to translate a WAVES port integer into a **Std_Logic** bit or a list of WAVES port integers into a **Std_Logic_Vector**. Only resolved **Std_Logic** data types are required, since this function was designed for bi-directional signals which require a resolved **Std_Logic** data type. Columns 1 and 5 of Table 6-1 show the mapping relationships for the WAVES logic values as we described in Section 4.3.1, as well as the standard logic simulation code that is produced from the WAVES port integer value.

The **Expect_1164** function also converts a single WAVES port integer into an 1164 **Std_Logic** value. The difference between **Expect_1164** and **Stim_1164** is that the function **Expect_1164** defines an expected or predicted value of a model's output. This generated signal is the definition of a predicted value which verifies conformance of the model's output. Therefore, for output pins, the **Expect_1164** function behaves the same as **Stim_1164**, except the logic value represents the expected response of the model. However, for bi-directional pins, the **Expect_1164** function strips off the driven values and replaces them with '-', the **Dont_Care** value. This translation of drive values to **Dont_Care** values allows the use of this function with any signal specified as OUT or INOUT. The translation function is also overloaded to translate a single port integer into a **Std_Logic** bit or a list of WAVES port integers into a **Std_Logic_Vector**. In this case, the resolved **Std_Logic_Vector** type is not required. Since this function was designed to translate the logic values for input into a VHDL process, that process requires an unresolved source for explicit definition of the expected model's output. The functionality of the VHDL monitor process that uses the translated value is described in more detail in Section 6.2.2. Columns 1 and 4 of Table 6-1 show the mapping relationships between the WAVES logic values, as described in Section 4.3.1, and the standard logic simulation code which is produced from the WAVES port integer value.

6.2.2 The WAVES Compatible Function

The **compatible** function evaluates the state of two 1164 **Std_Logic** bits or **Std_Logic_Vectors** for compatibility. This function was designed for use in the test-bench within a monitor process. It allows the process to be sensitive to the actual signal value from the model and the expected signal value from the WAVES dataset. The process utilizing this function determines if the two signals are "compatible" over the simulation period. A simple assertion statement can then be used to notify the designer when the two signals are *in*compatible. In this function, the actual data value is evaluated to determine if it is compatible with an expected or predicted value. By compatible, we mean: "Is the actual value equivalent to or equal to the value specified by the expected value?" The function uses a simple lookup table which contains all of the compatible definitions that were developed. For a more detailed analysis, the complete table is listed in Appendix D. The order of relationship between the actual and expected values must be preserved, since they carry different meanings. For example, if the actual data is '1' and the expected data was '-' (**Dont_Care**), the result would be true. However, if actual data is '-' and the expected data was '1', the result would be false. The function is overloaded to support all of the data types that would be utilized by any design compliant with IEEE STD 1164-1993.

Now, having described the WAVES utility package, and its integral translation functions and compare function, we are ready to look into the implementation of the simulation system.

6.3 Implementation of the WAVES-VHDL Simulation System

So far, we have presented all the necessary elements to implement a WAVES-VHDL simulation system or environment. We first described the purpose, background, and history of WAVES. Then we explained the waveform concepts, to provide a conceptual understanding of the WAVES and VHDL interface, and followed this with a description of the WAVES dataset representation of waveforms, along with the steps required to construct this dataset. Next, we presented certain library structure issues to support the WAVES-VHDL simulation environment. Finally, we introduced a set of library functions to assist us in developing the WAVES dataset and testbench.

Now it is time to put these elements together in order to simulate a WAVES-VHDL design. This implementation of the WAVES-VHDL simulation system consists of *three distinct logical steps*. They are *developing VHDL models for the UUT, generating the WAVES dataset, and generating the testbench.* We describe each step below, with appropriate examples.

6.3.1 Developing VHDL Models of the UUT

The purpose of any simulation is to verify the functionality or timing characteristics of a device or a system we wish to design. Hence, a necessary step before carrying out simulation is to capture the functionality and timing characteristics of the device or unit in VHDL. In general, there are three different ways to capture the device in VHDL. The first way is to describe a device by using a *behavioral architecture.* As the name implies, the behavioral architecture simply describes the device functionality without any hardware implementation details. A second way to describe the device is by using a *Register Transfer Level (RTL) architecture.* The RTL description implies some level of implementation details in the form of data flow. The third way to describe the device is by using a *structural architecture.* The structural description shows a full hardware implementation view of a design, similar to a netlist.

Then, given these three architecture options exist for a device entity, which architecture should be used to model the device? Actually, it depends on the accuracy and abstraction level desired and whether or not structural information is required. If the model is going to be used to drive a circuit layout tool, then the most adequate architecture would be the structural description. On the other hand, if structural information is not needed, and more efficient simulation performance is the goal, then one of the other two architectures will be more appropriate. Figure 6-6 shows an example of the behavioral description for a D-type flip-flop (**D_Flip_Flop**). If we wish to use this description, we create a work library within our VHDL environment first, and compile this entity and architecture pair into the work library.

```
        -- This is a behavioral model for a Positive edge
triggered D Flip Flop
              -- for a WAVES example.

              library ieee;
              use ieee.std_logic_1164.all;

              entity d_flip_flop is

                port ( clock : in  std_logic ;
                       D     : in  std_logic ;
                       Q     : out std_logic ;
                       Q_bar : out std_logic
                     );

              end d_flip_flop ;

              architecture behavioral of d_flip_flop is

              begin

              main : process ( clock )

                begin

                  if clock = '1' then
                    Q      <= D ;
                    Q_bar  <= not(D) ;

                  else
                    null;
                  end if;

                end process;

              end behavioral;
```

Figure 6-6. A Behavioral Architecture for a D-type Flip Flop

6.3.2 WAVES Dataset Generation

Our next step is to generate a WAVES dataset to stimulate our VHDL model (UUT) and provide the expected response to the monitor process, as we illustrated earlier in Figure 6-1. The WAVES dataset development procedure was presented in Section 5.2, and an improved and simplified WAVES dataset development procedure utilizing the library **WAVES_1164** was presented in Section 5.3. As shown earlier in Figure 5-11, the WAVES dataset generation process requires only four sub-steps, if we utilize the library **WAVES_1164**. In addition, WAVES requires a header file which contains information about the WAVES dataset. Next, we review the WAVES header file and the four sub-steps in the context of our D-type flip-flop example. For more specific information, we should consult Chapter 5.

6.3.2.1 Create the WAVES Header File

First, we need to create a header file for the dataset. Recall that the WAVES header file is provided for information only. It offers information about the WAVES dataset, including how it is structured and analyzed. If more information about the header file is needed, please refer to Section 4.2. Figure 6-7 illustrates the WAVES header file for the D-type flip-flop example.

```
-- **************************************************
--
-- ******** Header File for Entity: d_flip_flop
--
-- **************************************************
-- **************************************************
--
-- Data Set Identification Information
--
TITLE          A General Description
DEVICE_ID      d_flip_flop

DATE           Wed Sep  6 15:09:10 1995
ORIGIN         Company X Design Team
AUTHOR         Company or Person
AUTHOR         Maybe Multiple ... Companies or People
DATE           Wed Sep  6 15:09:10 1995
ORIGIN         Modified by Company X Design Team
AUTHOR         Who did it Company or Person

OTHER          Any general comments you want
OTHER          Built Using the WAVES-VHDL 1164 STD Libraries
--
```

```
-- Data Set Construction Information
--
WAVES_FILENAME     ./waves_pins.vhd                                WORK
library            WAVES_1164;
use                WAVES_1164.WAVES_1164_Pin_Codes.all;
use                WAVES_1164.WAVES_1164_Logic_Value.all;
use                WAVES_1164.WAVES_Interface.all;
use                WORK.UUT_Test_pins.all;
WAVES_UNIT         WAVES_OBJECTS                                   WORK
WAVES_FILENAME     ./waves_wgen.vhd                                WORK
--
EXTERNAL_FILENAME     vectors.txt                  VECTORS
--
WAVEFORM_GENERATOR_PROCEDURE     WORK.waves_d_flip_flop.waveform
```

Figure 6-7. The WAVES header file for the D-type Flip-Flop

6.3.2.2 Define and Compile UUT Test Pins Package

Using the **WAVES_1164** library, the first step in building a simulatable waveform is to define test pins that name the pins to which the waveform applies. This package needs to be analyzed prior to compilation of the **Waves_Object** package, which requires visibility of the test pins package. Figure 6-8 shows an example of the test pins specification for the D-type flip-flop. This package must be compiled into the work library.

```
-- WAVES Test Pins Specification

PACKAGE uut_test_pins IS
--  Define UUT's Test Pins
       TYPE test_pins IS (clock, D, Q, Q_BAR);
END uut_test_pins;
--
```

Figure 6-8. Test Pins for the D Flip-Flop

6.3.2.3 Compile the WAVES Objects Package

Next, the **Waves_Objects** package needs to be compiled into the work library. The **Waves_Objects** package is dependent on the private types declared in the **Waves_System** package. It also requires visibility of TEXTIO, **Waves_Interface**, **Waves_1164_Logic_Value**, **Waves_1164_Pin_Codes**, and the

UUT **Test_Pins** packages. The context clause shown in Figure 6-9 needs to be added to the beginning of the **Waves_Objects** package prior to its compilation. Even though the **Waves_Objects** package is one of the WAVES standard packages, it requires visibility of the **Test_Pins** which is a device-specific package, and it needs to be visible to the waveform generator procedure which is also device-specific. Hence, this package needs to be compiled into the default work library for each UUT.

```
--
use STD.TEXTIO.all;
--  Object package use TEXTIO
library WAVES_STD;
use     WAVES_STD.WAVES_SYSTEM;
Library WAVES_1164;
use WAVES_1164.WAVES_Interface.all
use WAVES_1164.WAVES_1164_Logic_Value.all;
use WAVES_1164.WAVES_1164_Pin_Codes.all;
use Work.UUT_Test_pins.all;
--
```

Figure 6-9. Context Clause of the Waves_Object Package

6.3.2.4 Define and Compile the Waveform Generator Procedure Package

The next step in building a waveform is to define one or more waveform generator procedures (WGPs). A waveform generator procedure is executed to generate a waveform. This package requires visibility of the **Waves_Standard**, TEXTIO, **Waves_1164_Frames**, **Waves_1164_Pin_Codes**, **Waves_Interface**, **Waves_Objects**, and **Test_Pins** packages. Figure 6-10 shows a complete example of the WGP for the D-type flip-flop, including the context clauses. This package needs to be compiled into the work library since it is a device-specific package.

```
LIBRARY WAVES_STD;
USE WAVES_STD.WAVES_Standard.all;

USE STD.textio.all;
LIBRARY WAVES_1164;
USE WAVES_1164.waves_1164_frames.all;
USE WAVES_1164.waves_1164_pin_codes.all;
USE WAVES_1164.waves_interface.all;
USE work.waves_objects.all;
USE work.uut_test_Pins.all;

PACKAGE WGP_d_flip_flop is
    PROCEDURE  waveform(SIGNAL WPL : inout WAVES_PORT_LIST);
END WGP_d_flip_flop;
```

```
-------------------------------------------------------

PACKAGE BODY WGP_d_flip_flop is

-- This is the uut pin declaration pin and ordering
-- Remember you need to match the External file to This order
--
--clock, D, Q, Q_bar

    PROCEDURE  waveform(SIGNAL WPL : inout WAVES_PORT_LIST) is

       FILE vector_file : TEXT is in "vector.txt";

       VARIABLE vector : FILE_SLICE := NEW_FILE_SLICE;

       -- declare time constants to use or use time literals
       -- constants or time literals can be used as the
       -- frame time values

         CONSTANT outputs: pinset:= new_pinset((Q, Q_bar));
         CONSTANT inputs: pinset:= new_pinset((D));

         CONSTANT vector_FSA : Frame_set_array :=
          New_frame_set_array(Pulse_high( 5 ns, 15 ns), clock) +
          New_frame_set_array(Non_return( 0 ns), inputs) +
          New_frame_set_array(window( 10 ns, 20 ns), outputs);

       VARIABLE timing : time_data := new_time_data(vector_fsa);

      BEGIN
        loop
          READ_FILE_SLICE (vector_file, Vector);  -- get first
          vector
          exit when vector.end_of_file;
          apply(wpl, vector.codes.all, Delay(vector.fs_time),
          timing);
        end loop;
END waveform;

END WGP_d_flip_flop;
```

Figure 6-10. Waveform Generator Procedure for the D-type Flip-Flop

6.3.2.5 Create the WAVES External File

Our final step in creating the dataset is to implement an external file. The external file is a pure data or text file which does not need to be analyzed and it is only referenced by the waveform generator procedure during simulation. Hence, it can be created anytime prior to simulation of the WGP, which means that changing the external file does not affect the compilation of WAVES files. When we create an external file, the order of pin codes for each signal or pin in the file (columns in external file) must be the same order as the test pins declaration. The "%" character denotes the beginning of a comment line. Figure 6-11 shows a test vector file (an external file) for the D-type Flip-Flop.

```
%clock D   Q   QBAR
     1   1    1    0 : 20 ns;
     0   0    1    0 : 20 ns;
     1   0    0    1 : 20 ns;
     1   1    1    0 : 20 ns;
     1   0    0    1 : 20 ns;
```

Figure 6-11. Test vectors for the D-type Flip-Flop

6.3.3 Testbench Generation

After we develop the VHDL model for the UUT and the WAVES dataset, the final step to carry out is the testbench generation. It is the testbench which applies a stimulus to the model and checks the model's responses, as we illustrated earlier in Figure 6-1. The testbench is a free-running model which will execute for the time duration specified or until the external vector file is exhausted. During simulation, the testbench will report the success or failure of the simulation. Figure 6-12 shows the testbench for the D-type flip-flop example. Here we utilize the testbench utility functions described in Section 6.2 to facilitate the testbench generation.

```
--
LIBRARY ieee;
USE ieee.std_logic_1164.ALL;

LIBRARY waves_1164;
USE waves_1164.WAVES_1164_utilities.all;

USE WORK.UUT_test_pins.all;
USE work.waves_objects.all;
USE work.WGP_d_flip_flop.all;

-- User Must Modify And ADD resource library references here
```

```
ENTITY test_bench IS
END test_bench;

ARCHITECTURE d_flip_flop_test OF test_bench IS

   COMPONENT d_flip_flop
    PORT ( clock              : IN   std_logic;
           D                  : IN   std_logic;
           Q                  : OUT  std_logic;
           Q_bar              : OUT  std_logic);
     END COMPONENT;

  --*****************************************************
  --***********CONFIGURATION SPECIFICATION ***********
  --*****************************************************
 -- Modify entity use statement
 --   .. Architecture, Library, Component ..

FOR ALL:d_flip_flop USE ENTITY work.d_flip_flop(behavioral);

  --**********************************************************
  -- stimulus signals for the waveforms mapped into UUT INPUTS
  --**********************************************************

    SIGNAL WAV_STIM_clock                   :std_logic;
    SIGNAL WAV_STIM_D                       :std_logic;

  --****************************************************
  -- Expected signals used in monitoring the UUT OUTPUTS
  --****************************************************

    SIGNAL FAIL_SIGNAL                      : std_logic;
    SIGNAL WAV_EXPECT_Q                       :std_logic;
    SIGNAL WAV_EXPECT_Q_bar                   :std_logic;

  --*******************************************************
  -- UUT Output signals used In Monitoring ACTUAL Values
  --*******************************************************

    SIGNAL ACTUAL_Q                       :std_logic;
    SIGNAL ACTUAL_Q_bar                   :std_logic;

--************************************************************
-- WAVES signals OUTPUTing each slice of the waves port list
--************************************************************

        SIGNAL wpl  : WAVES_port_list;
```

```
BEGIN
  --
  --*********************************************************************
  -- process that generates the WAVES waveform
  --*********************************************************************

      WAVES: waveform(wpl);

  --********************************************************************
  -- processes that convert the WPL values to 1164 Logic Values
  --********************************************************************

  WAV_STIM_clock              <= STIM_1164(wpl.wpl( 1 ));
  WAV_STIM_D                  <= STIM_1164(wpl.wpl( 2 ));
  WAV_EXPECT_Q                  <= EXPECT_1164(wpl.wpl( 3 ));
  WAV_EXPECT_Q_bar              <= EXPECT_1164(wpl.wpl( 4 ));

  --******************************************
  -- UUT Port Map - Name Symantics Denote Usage
  --******************************************

  u1: d_flip_flop
  PORT MAP(
    clock                 => WAV_STIM_clock,
    D                     => WAV_STIM_D,
    Q                     => ACTUAL_Q,
    Q_bar                 => ACTUAL_Q_bar);


  --*******************************************************************
  -- Monitor Processes To Verify The UUT Operational Response
  --*******************************************************************

Monitor_Q:
  PROCESS(ACTUAL_Q, WAV_expect_Q)
  BEGIN
       assert(Compatible (actual => ACTUAL_Q,
                          expected => WAV_expect_Q))
       report "Error on Q output" severity WARNING;

  IF ( Compatible ( ACTUAL_Q,    WAV_expect_Q) ) THEN
    FAIL_SIGNAL <='L'; ELSE FAIL_SIGNAL <='1';
  END IF;
  END PROCESS;

Monitor_Q_bar:
  PROCESS(ACTUAL_Q_bar, WAV_expect_Q_bar)
```

```
   BEGIN
        assert(Compatible (actual => ACTUAL_Q_bar,
                           expected => WAV_expect_Q_bar))
        report "Error on Q_bar output" severity WARNING;

   IF ( Compatible ( ACTUAL_Q_bar,    WAV_expect_Q_bar) ) THEN
     FAIL_SIGNAL <='L'; ELSE FAIL_SIGNAL <='1';
   END IF;
   END PROCESS;

END d_flip_flop_test;
```

Figure 6-12. The Testbench for D Flip-Flop

There are a number of logical elements in the testbench for which we must ensure both presence and correctness. They are context clauses, component and configuration declarations, connection signals, invocation of the waveform generator, translation functions, UUT instantiation, and monitor processes. In the following sections we describe each of these elements in the testbench. By the way, all elements in the testbench run concurrently. Therefore, the order in which each element appears in the architecture will not influence the simulation results. For instance, let's say that the first element in the architecture is the waveform generator and the last element is the monitor processes. Even if we change these two elements' positions in the simulation, our results will remain the same.

6.3.3.1 Context Clauses

The testbench requires visibility of many packages in various libraries. When the testbench is developed utilizing IEEE Standard 1164 logic, visibility of the **Std_Logic_1164** package in the IEEE library is needed. The testbench also utilizes translation and compare functions in the **Waves_1164_Utilities** package which resides in the library **WAVES_1164**. In addition, packages that define **Test_Pins**, **Waves_Objects**, and waveform generator procedures need to be visible to the testbench because they are all utilized by it. Adequate context clauses that ensure the necessary visibility for the D-type flip-flop example are presented below.

```
LIBRARY ieee;
USE ieee.std_logic_1164.ALL;

LIBRARY waves_1164;
USE waves_1164.WAVES_1164_utilities.all;

USE WORK.UUT_test_pins.all;
USE work.waves_objects.all;
```

```
USE work.WGP_d_flip_flop.all;

-- If testbench instantiates other components
--   in resource library

-- user must modify and add an appropriate library
--   and use clauses here.
```

6.3.3.2 Component and Configuration Declarations

According to VHDL definition, a component must be declared before it is instantiated in the architecture, and all components must be configured by using configuration statements. There are some variations of the configuration statement. However, the language definition must be followed. We may wish to consult IEEE STD 1076-1993 (or 1987) for more information about the configuration statement. An example of a component and a corresponding configuration statement, for the D-type flip-flop, are shown below. Here, we have configured an entity **D_Flip_Flop** with a behavioral architecture.

```
COMPONENT d_flip_flop
 PORT ( clock                    : IN   std_logic;
        D                        : IN   std_logic;
        Q                        : OUT  std_logic;
        Q_bar                    : OUT  std_logic);
  END COMPONENT;

--*******************************************************
--***********CONFIGURATION SPECIFICATION ***********
--*******************************************************

-- User Must  modify and declare correct
--    .. Architecture, Library, Component ..

FOR ALL:d_flip_flop USE ENTITY work.d_flip_flop(behavioral);
```

6.3.3.3 Connection Signals

Next, an appropriate internal connection signal set needs to be declared. These connection signals are used to connect all processes and components in the testbench. We present an example of the connection signals for the D-type flip-flop below. In this example, the signal **Fail_Signal** is used to depict the status of the test in the monitor processes. If the expected and actual response do not match, this signal is set to '1'; otherwise, it is set to 'L'.

```
--**************************************************************
-- stimulus signals for the waveforms mapped into UUT INPUTS
--**************************************************************
  SIGNAL WAV_STIM_clock                     :std_logic;
  SIGNAL WAV_STIM_D                         :std_logic;

--*******************************************************
-- Expected signals used in monitoring the UUT OUTPUTS
--*******************************************************
  SIGNAL FAIL_SIGNAL                        : std_logic;
  SIGNAL WAV_EXPECT_Q                         :std_logic;
  SIGNAL WAV_EXPECT_Q_bar                     :std_logic;

--*********************************************************
-- UUT Output signals used In Monitoring ACTUAL Values
--*********************************************************
  SIGNAL ACTUAL_Q                         :std_logic;
  SIGNAL ACTUAL_Q_bar                     :std_logic;
```

6.3.3.4 Invocation of the Waveform Generator

The testbench invokes the waveform generator procedure to stimulate the UUT and to provide the expected response of the UUT to the monitor processes. The waveform generator procedure sends both the stimulus and the expected response through a **Waves_Port_List** signal type. Hence, a signal of type **Waves_Port_List** must be defined for the waveform generator. The type **Waves_Port_List** is defined in the **Waves_Objects** package. The following shows an example of the signal declaration and invocation of the waveform generator for the D-type flip-flop.

```
--**************************************************************
-- WAVES signals OUTPUTing each slice of the waves port list
--**************************************************************

     SIGNAL wpl   : WAVES_port_list;

--*************************************************************
-- process that generates the WAVES waveform
--*************************************************************

     WAVES: waveform(wpl);
```

6.3.3.5 Translation Functions

As we mentioned in Section 6.2, the waveform generator procedure generates integer, instead of the standard logic 1164, values for the stimulus data and expected responses on the **Waves_Port_List** signal. Hence, the integer values on the **Waves_Port_List** need to be translated into the standard logic 1164 type, to interface with the UUT and monitor processes. We accomplish these translations by utilizing functions in the **Waves_1164_Utilities** package. The following example demonstrates the use of translation functions for the D-type flip-flop example. In particular, the functions **Stim_1164** and **Expect_1164** are used in this example.

```
--*************************************************************
-- processes that convert the WPL values to 1164 Logic Values
--*************************************************************

WAV_STIM_clock                <= STIM_1164(wpl.wpl( 1 ));
WAV_STIM_D                    <= STIM_1164(wpl.wpl( 2 ));
WAV_EXPECT_Q                  <= EXPECT_1164(wpl.wpl( 3 ));
WAV_EXPECT_Q_bar              <= EXPECT_1164(wpl.wpl( 4 ));
```

6.3.3.6 UUT Instantiation

Next, the testbench instantiates the VHDL model(s) (the UUTs). The testbench stimulates the UUT with the translated stimulus data from the waveform generator, and then the UUT provides actual responses to the monitor processes. We illustrate the instantiation of the D-type flip-flop example below.

```
--********************************************
-- UUT Port Map - Name Symantics Denote Usage
--********************************************

u1: d_flip_flop
PORT MAP(
  clock              => WAV_STIM_clock,
  D                  => WAV_STIM_D,
  Q                  => ACTUAL_Q,
  Q_bar              => ACTUAL_Q_bar);
```

6.3.3.7 Monitor Processes

Finally, the testbench invokes the monitor processes for each output of the UUT. The monitor processes compare the actual response of the UUT with the expected response from the waveform generator. A monitor process utilizes an overloaded compare function (Compatible function in **Waves_1164_Utilities** package, see Section 6.2) to verify the functionality of the UUT during simulation. The monitor processes for the D-type flip-flop are presented below.

```
--*************************************************************
-- Monitor Processes To Verify The UUT Operational Response
--*************************************************************

Monitor_Q:
  PROCESS(ACTUAL_Q, WAV_expect_Q)
  BEGIN
       assert(Compatible (actual => ACTUAL_Q,
                          expected => WAV_expect_Q))
       report "Error on Q output" severity WARNING;

  IF ( Compatible ( ACTUAL_Q,    WAV_expect_Q) ) THEN
    FAIL_SIGNAL <='L'; ELSE FAIL_SIGNAL <='1';
  END IF;
  END PROCESS;

Monitor_Q_bar:
  PROCESS(ACTUAL_Q_bar, WAV_expect_Q_bar)
  BEGIN
       assert(Compatible (actual => ACTUAL_Q_bar,
                          expected => WAV_expect_Q_bar))
       report "Error on Q_bar output" severity WARNING;

  IF ( Compatible ( ACTUAL_Q_bar,    WAV_expect_Q_bar) ) THEN
    FAIL_SIGNAL <='L'; ELSE FAIL_SIGNAL <='1';
  END IF;
  END PROCESS;
```

In ***summary of our Implementation of the WAVES-VHDL Simulation System,*** we presented the three steps necessary to implement a WAVES-VHDL simulation system. They are: (1) developing VHDL models for the UUT, (2) generating the WAVES dataset, and (3) generating the testbench. As we described above, each step in turn consists of several sub-steps. With the support of the WAVES_STD and WAVES_1164 libraries, we need to specify only the device-dependent data in the work library for simulation, as we illustrate in Figure 6-13. To demonstrate this, we compiled all of the WAVES files and VHDL files we have created or used in our D-type flip-flop example, and executed a simulation. The simulation results are shown in Figure 6-14. All the source code files for this example are included on the companion CD-ROM in a directory called "FlipFlop".

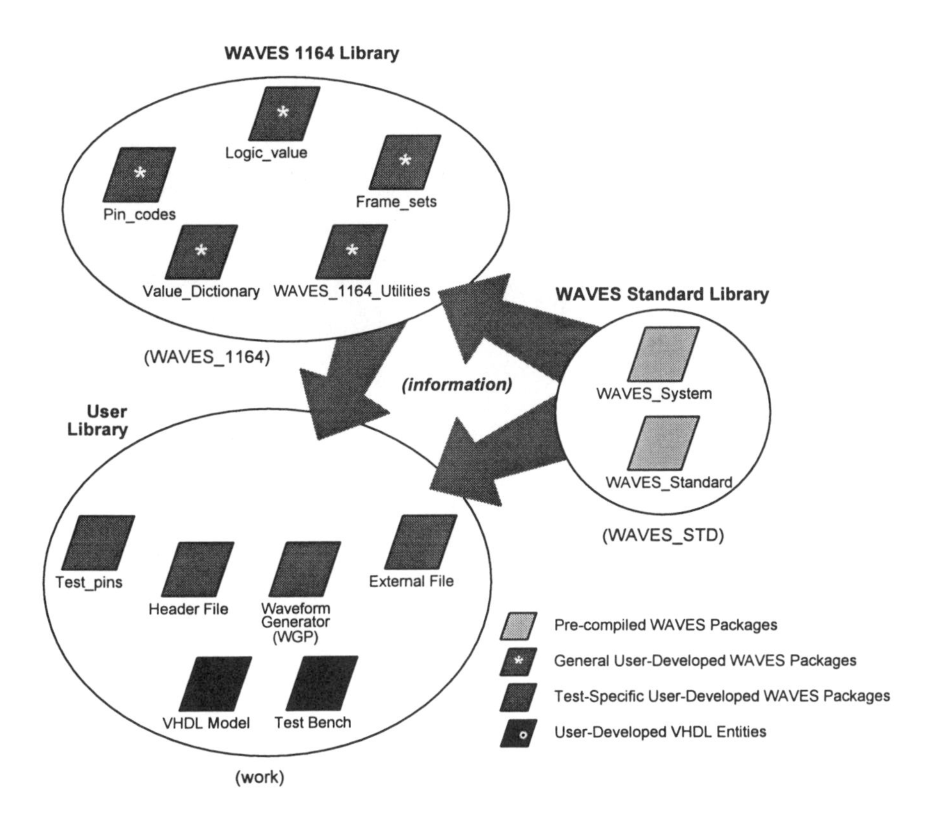

Figure 6-13. Library support for a WAVES-VHDL Simulation

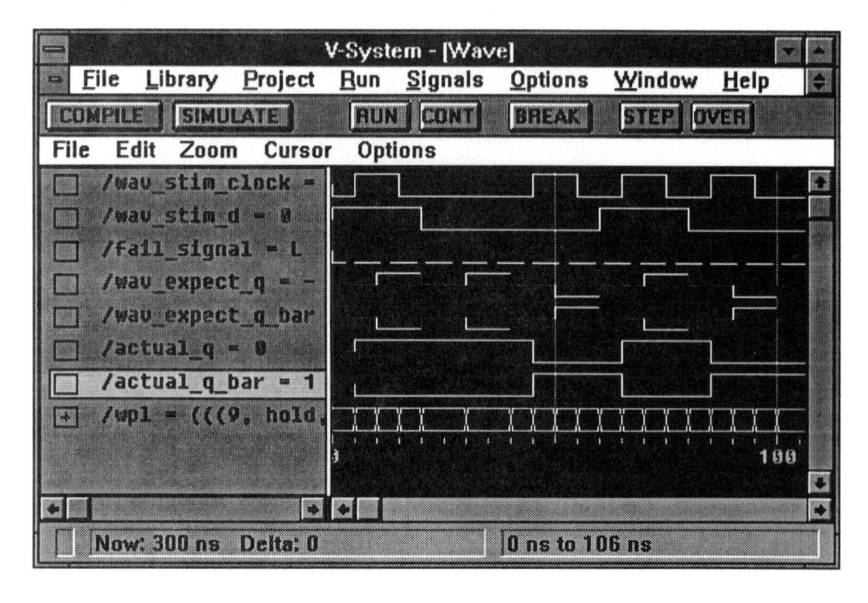

Figure 6-14. Simulation results of the D Flip-Flop

Now, having described the WAVES-VHDL simulation implementation, and completed our D-type flip-flop example, we are ready to apply the WAVES process to another case. In the next section, we undertake a second example.

6.4 Example: The 54/74180 8-Bit Parity Generator/Checker

In this section, we develop another example, the 8-Bit Parity Generator/Checker, to further demonstrate the use of WAVES and VHDL in a design and test environment. First, we provide device specifications which describe the functionality and input/output interfaces. Next, we develop all the necessary files (as we did in Section 6.3) to support simulation of the WAVES-VHDL design. Finally, we present the simulation results for this second example.

6.4.1 Device Specifications

This parity checker was modeled after the 54/74180 series taken from the Fairchild TTL Data Book printed in December 1978. The '180 is a monolithic, 8-bit parity checker/generator which features control inputs and even or odd outputs to enhance operation in either odd or even parity applications. Cascading these circuits allows unlimited word length expansion. A typical application would be to generate and check parity on data being transmitted from one register to another. It is a 14 pin device with 8 data inputs (I0 through I7), 2 control inputs (Odd Input and Even Input) and 2 outputs (Odd Parity Output and Even Parity Output). Figure 6-15 lists the pin numbers, the pin names, and a short description of each pin. Figure 6-16 is the truth table for the device.

Pin #	Pin Name	Description
1	I6	Data Input 6
2	I7	Data Input 7
3	EI	Even Input
4	OI	Odd Input
5	SE	Even Parity Output
6	SO	Odd Parity Output
7	GND	Ground
8	I0	Data Input 0
9	I1	Data Input 1
10	I2	Data Input 2
11	I3	Data Input 3
12	I4	Data Input 4
13	I5	Data Input 5
14	Vcc	Supply Voltage

Figure 6-15. 8-Bit Parity Checker/Generator Pin Labels.

Inputs			Outputs	
Parity	Even	Odd	Even	Odd
Even	H	L	H	L
Odd	H	L	L	H
Even	L	H	L	H
Odd	L	H	H	L
X	H	H	L	L
X	L	L	H	H

Figure 6-16. 8-Bit Parity Checker/Generator Truth Table.

6.4.2 VHDL Models for the UUT, WAVES Dataset, and Testbench

As we explained in Section 6.3, the first step toward simulation of a WAVES-VHDL design is to develop a VHDL model for the UUT. Once the VHDL model (entity and architecture) of the UUT is completed, the necessary WAVES files needed to test or stimulate the UUT are the header file, test pins file, objects file, waveform generator file, and the vector file. Finally, a testbench needs to be created for the simulation. We developed all of these files for our example, and they are presented on the following pages.

6.4.2.1 A VHDL Model for the 54/74180 (UUT)

We developed a behavioral description of the 8-Bit Parity Generator/Checker, and it is presented in Figure 6-17 below. We added some comments to capture certain useful information such as model identification, author, organization, platform, and the model revision history. These comments are not required; however, good engineering practice dictates that we should provide an adequate level of comments for documentation purposes.

```
-- TITLE: Fairchild 54/74180 8-BIT PARITY GENERATOR/CHECKER
-- DATE : xx yyy xxx
--
-- VERSION : x.x
-- FILENAME: parity_gen.vhd
-- FUNCTION: Entity, Architecture for the Fairchild 54/74180
--           8-BIT PARITY GENERATOR/CHECKER - Behavioral Model.
--
-- AUTHOR       : xxxxx xxxx
-- ORGANIZATION: YYYY YYYYY
--
-- PURPOSE AND USE: This was written as an example to
   demonstrate
--                  WAVES usage.
--
-- TIMING: None


-- NOTES :
--
-- DEVELOPMENT PLATFORM : Pentium w/WIN'95
-- VHDL SOFTWARE VERSION: MTI VSYSTEM/WINDOWS v4.2f
--
-- Rev.  HISTORY:
--xx yyy 95 - v2.0 - Includes timing
-- 9 May 95 - v1.5 - Changed to case statement, still no timing.
-- 5 May 95 - v1.0 - Initial version, functional only, no
  timing.
--
--
```

```
library ieee;
use ieee.std_logic_1164.all;

ENTITY parity_generator_checker IS
PORT(  Data_input           :IN    STD_LOGIC_VECTOR(0 TO 7);
       Odd_input            :IN    STD_LOGIC;
       Even_input           :IN    STD_LOGIC;
       Odd_parity_output    :OUT   STD_LOGIC;
       Even_parity_output   :OUT   STD_LOGIC);
END parity_generator_checker;

ARCHITECTURE behavioral OF parity_generator_checker IS

BEGIN

PROCESS(Data_input, Odd_input, Even_input)

variable parity             :STD_LOGIC;
variable controls    :STD_LOGIC_VECTOR(0 TO 2);
variable tplh_odd    :TIME;
variable tphl_odd    :TIME;
variable tplh_even   :TIME;
variable tphl_even   :TIME;

BEGIN

-- Generate Timing

IF Odd_input = '0' AND Data_input'EVENT THEN
       tplh_odd := 48 NS;
       tphl_odd := 38 NS;
       tplh_even := 60 NS;
       tphl_even := 68 NS;
END IF;
IF Even_input = '0' AND Data_input'EVENT THEN
       tplh_odd := 60 NS;
       tphl_odd := 68 NS;
       tplh_even := 48 NS;
       tphl_even := 38 NS;
END IF;
IF NOT Data_input'EVENT AND
           (Odd_input'EVENT OR Even_input'EVENT) THEN
       tplh_odd := 20 NS;
       tphl_odd := 10 NS;
       tplh_even := 20 NS;
       tphl_even := 10 NS;
END IF;
```

```
-- Calculate Parity

parity := '0';

FOR x IN Data_input'RANGE LOOP
  parity := parity XOR Data_input(x);
END LOOP;


-- Assign Output
controls := parity & Odd_input & Even_input;

CASE controls IS
WHEN "100" | "000" =>
        Odd_parity_output <= '1' AFTER tplh_odd;
        Even_parity_output <= '1' AFTER tplh_even;
WHEN "101" | "010" =>
       Odd_parity_output <= '1' AFTER tplh_odd;
       Even_parity_output <= '0' AFTER tphl_even;
WHEN "110" | "001" =>
       Odd_parity_output <= '0' AFTER tphl_odd;
       Even_parity_output <= '1' AFTER tplh_even;
WHEN "111" | "011" =>
       Odd_parity_output <= '0' AFTER tphl_odd;
       Even_parity_output <= '0' AFTER tphl_even;
WHEN others =>
        null;
END CASE;

END PROCESS;
END behavioral;
```

Figure 6-17. A Behavioral Model for the 54/74180

6.4.2.2 The WAVES Dataset

In this section, we present all of the WAVES files that make up WAVES dataset for the 8-bit Parity Checker/Generator.

6.4.2.2.1 The Header File

The WAVES header file provides information about the WAVES dataset, including how it is structured and analyzed. The header file for the 8-bit Parity Checker/Generator is provided in Figure 6-18 below.

```
-- ***************************************************
--
-- ******** Header File for Entity: parity_generator_checker
--
-- ***************************************************
-- ***************************************************
--
-- Data Set Identification Information
--
TITLE          A General Description
DEVICE_ID      parity_generator_checker

DATE           Thu Sep  7 09:36:56 1995
ORIGIN         Company X Design Team
AUTHOR         Company or Person
AUTHOR         Maybe Multiple ... Companies or People
DATE           Thu Sep  7 09:36:56 1995
ORIGIN         Modified by Company X Design Team
AUTHOR         Who did it Company or Person

OTHER          Any general comments you want
OTHER          Built Using the WAVES-VHDL 1164 STD Libraries
--
-- Data Set Construction Information
--
WAVES_FILENAME   ./waves_pins.vhd                                WORK
library          WAVES_1164;
use              WAVES_1164.WAVES_1164_Pin_Codes.all;
use              WAVES_1164.WAVES_1164_Logic_Value.all;
use              WAVES_1164.WAVES_Interface.all;
use              WORK.UUT_Test_pins.all;
WAVES_UNIT       WAVES_OBJECTS                                   WORK
WAVES_FILENAME   ./waves_wgen.vhd                                WORK
--
EXTERNAL_FILENAME vectors.txt  VECTORS
--
WAVEFORM_GENERATOR_PROCEDURE
WORK.waves_parity_generator_checker.waveform
```

Figure 6-18. The header file for 54/74180

6.4.2.2.2 The Test Pins Package

The type **Test_Pins** lists all the input and output pins which will be used to define pin sets for the waveform generator. The test pins for the 8-bit Parity Checker/Generator are provided in Figure 6-19.

```
-- ******** Generated for Entity: parity_generator_checker
-- ******** This File Was Generated on: Thu Sep  7 09:36:56 1995
--
--
PACKAGE uut_test_pins IS
TYPE test_pins IS (Data_input_0, Data_input_1, Data_input_2,
                    Data_input_3, Data_input_4, Data_input_5,
                   Data_input_6, Data_input_7, Odd_input,
                    Even_input, Odd_parity_output,
                    Even_parity_output);
END uut_test_pins;
```

Figure 6-19. Test Pins for the 54/74180

6.4.2.2.3 The Waves_Objects Package

The **Waves_Objects** package developed by TASG of the IEEE is provided on the companion CD-ROM. However, this package needs to be modified to ensure the visibility requirements of VHDL. Here, we present only the modifications needed at the beginning of the **Waves_Objects** package. In particular, the context clauses that need to be added are presented in Figure 6-20 below. These context clauses must be added to the beginning of the **Waves_Objects** package.

```
--
--
use STD.TEXTIO.all;
library WAVES_STD;
use WAVES_STD.WAVES_SYSTEM;
-- A context clause providing visibility to an
-- analyzed  copy of WAVES_INTERFACE is required
-- at this point. Context clauses providing
-- visibility to LOGIC VALUE, TEST PINS,
-- and PIN CODES are required at this point.
Library WAVES_1164;
use WAVES_1164.WAVES_Interface.all;
use WAVES_1164.WAVES_1164_Logic_Value.all;
```

```
use WAVES_1164.WAVES_1164_Pin_Codes.all;
use WORK.UUT_Test_pins.all;

package WAVES_OBJECTS is
        . . .
        . . .
        . . .
```

Figure 6-20. The Context Clauses for the Waves_Objects package

6.4.2.2.4 The Waveform Generator Procedure

The waveform generator procedure constructs a waveform by reading an external vector file. It provides both the stimulus and the expected response of the UUT to the testbench. Figure 6-21 illustrates a waveform generator procedure for the 8-bit parity generator/checker example.

```
LIBRARY WAVES_STD;
USE WAVES_STD.WAVES_Standard.all;

USE STD.textio.all;
LIBRARY WAVES_1164;
USE WAVES_1164.waves_1164_frames.all;
USE WAVES_1164.waves_1164_pin_codes.all;
USE WAVES_1164.waves_interface.all;
USE work.waves_objects.all;
USE work.uut_test_Pins.all;

PACKAGE WGP_parity_generator_checker is
     PROCEDURE  waveform(SIGNAL WPL : inout WAVES_PORT_LIST);
END WGP_parity_generator_checker;

-------------------------------------------------------

PACKAGE BODY WGP_parity_generator_checker is

-- This is the uut pin declaration pin and ordering
-- Remember you need to match the External file to This order
--
--Data_input_0, Data_input_1, Data_input_2, Data_input_3,
--Data_input_4, Data_input_5, Data_input_6, Data_input_7,
--Odd_input, Even_input, Odd_parity_output, Even_parity_output

    PROCEDURE  waveform(SIGNAL WPL : inout WAVES_PORT_LIST) is
```

```
        FILE vector_file : TEXT is in "vectors.txt";

        VARIABLE vector : FILE_SLICE := NEW_FILE_SLICE;

        CONSTANT Data_input: pinset:= new_pinset(( Data_input_0,
                                        Data_input_1,Data_input_2,
                                        Data_input_3,Data_input_4,
                                        Data_input_5,Data_input_6,
                                        Data_input_7));

        CONSTANT outputs: pinset:= new_pinset((Odd_parity_output,
                                             Even_parity_output));

        CONSTANT in_pins: pinset:= new_pinset((odd_input,
                                                   even_input));

        CONSTANT inputs: pinset:= in_pins or Data_input;


        CONSTANT vector_FSA : Frame_set_array :=
            New_frame_set_array(Non_return( 0 ns), inputs) +
            New_frame_set_array(window( 70 ns, 100 ns), outputs);

        VARIABLE timing : time_data := new_time_data(vector_fsa);

       BEGIN
         loop
          -- get first vector
           READ_FILE_SLICE (vector_file, Vector);
           exit when vector.end_of_file;
           apply(wpl, vector.codes.all,
                      Delay(vector.fs_time), timing);
         end loop;

       END waveform;

END WGP_parity_generator_checker;
```

Figure 6-21. Waveform Generator Procedure for the 54/74180

6.4.2.2.5 The External Test Vector File

The external file contains test vectors that specify a waveform in conjunction with frame set arrays. Figure 6-22 shows the test vector file (the external file) generated to test the 8-bit parity generator/checker.

```
%DDDD DDDD     e
%aaaa aaaa  e ov
%tttt tttt ov de
%aaaa aaaa de dn
%|||| |||| dn ||
%iiii iiii || oo
%nnnn nnnn ii uu
%pppp pppp nn tt
%uuuu uuuu pp pp
%tttt tttt uu uu
%0123 4567 tt tt

 0000 0000 01 01 : 100 NS;
 0000 0001 01 10 : 100 NS;
 0000 0010 01 10 : 100 NS;
 0000 0011 01 01 : 100 NS;
 0000 0000 10 10 : 100 NS;
 0000 0001 10 01 : 100 NS;
 0000 0010 10 01 : 100 NS;
 0000 0011 10 10 : 100 NS;
 0000 0000 00 11 : 100 NS;
 0000 0001 00 11 : 100 NS;
 0000 0010 00 11 : 100 NS;
 0000 0011 00 11 : 100 NS;
 0000 0000 11 00 : 100 NS;
 0000 0001 11 00 : 100 NS;
 0000 0010 11 00 : 100 NS;
 0000 0011 11 00 : 100 NS;
```

Figure 6-22. External file for the 54/74180

6.4.2.3 The Testbench

It is the testbench which applies the stimulus to the model (UUT) and checks the model's responses. We described the testbench in great detail in Section 6.3.3. Here, in Figure 6-23, we present a testbench that tests the 8-bit parity generator/checker.

```
 --
LIBRARY ieee;
USE ieee.std_logic_1164.ALL;

LIBRARY waves_1164;
USE waves_1164.WAVES_1164_utilities.all;

USE WORK.UUT_test_pins.all;
USE work.waves_objects.all;

USE work.WGP_parity_generator_checker.all;
-- Include component libary references here if applicable

ENTITY test_bench IS
END test_bench;

ARCHITECTURE parity_generator_checker_test OF test_bench IS

  --*******************************************************
  --***********CONFIGURATION SPECIFICATION ***************
  --*******************************************************

   COMPONENT parity_generator_checker
    PORT ( Data_input        : IN   std_logic_vector(  0 to  7 );
           Odd_input           : IN   std_logic;
           Even_input          : IN   std_logic;
           Odd_parity_output   : OUT  std_logic;
           Even_parity_output  : OUT  std_logic);
     END COMPONENT;

FOR ALL:parity_generator_checker USE ENTITY
work.parity_generator_checker(behavioral);

  --*************************************************************
  -- stimulus signals for the waveforms mapped into UUT INPUTS
  --*************************************************************

    SIGNAL WAV_STIM_Data_input    :std_logic_vector(  0 to  7 );
    SIGNAL WAV_STIM_Odd_input           :std_logic;
    SIGNAL WAV_STIM_Even_input          :std_logic;

  --*******************************************************
  -- Expected signals used in monitoring the UUT OUTPUTS
  --*******************************************************

    SIGNAL FAIL_SIGNAL                     : std_logic;
    SIGNAL WAV_EXPECT_Odd_parity_output    :std_logic;
```

```
    SIGNAL WAV_EXPECT_Even_parity_output  :std_logic;

  --********************************************************
  -- UUT Output signals used In Monitoring ACTUAL Values
  --********************************************************

    SIGNAL ACTUAL_Odd_parity_output   :std_logic;
    SIGNAL ACTUAL_Even_parity_output  :std_logic;


  --************************************************************
  -- WAVES signals OUTPUTing each slice of the waves port list
  --************************************************************

        SIGNAL wpl  : WAVES_port_list;

BEGIN
  --
  --*********************************************************
  -- process that generates the WAVES waveform
  --*********************************************************

       WAVES: waveform(wpl);

  --***********************************************************
  -- processes that convert the WPL values to 1164 Logic Values
  --***********************************************************

  WAV_STIM_Data_input            <= STIM_1164(wpl.wpl( 1 to 8 ));
  WAV_STIM_Odd_input             <= STIM_1164(wpl.wpl( 9 ));
  WAV_STIM_Even_input            <= STIM_1164(wpl.wpl( 10 ));
  WAV_EXPECT_Odd_parity_output    <= EXPECT_1164(wpl.wpl( 11 ));
  WAV_EXPECT_Even_parity_output   <= EXPECT_1164(wpl.wpl( 12 ));

  --******************************************
  -- UUT Port Map - Name Symantics Denote Usage
  --******************************************

  u1: parity_generator_checker
  PORT MAP(
    Data_input           => WAV_STIM_Data_input,
    Odd_input            => WAV_STIM_Odd_input,
    Even_input           => WAV_STIM_Even_input,
    Odd_parity_output    => ACTUAL_Odd_parity_output,
    Even_parity_output   => ACTUAL_Even_parity_output);
```

```
  --**********************************************************
  -- Monitor Processes To Verify The UUT Operational Response
  --**********************************************************

 Monitor_Odd_parity_output:
   PROCESS(ACTUAL_Odd_parity_output,
                 WAV_expect_Odd_parity_output)
   BEGIN
        assert(Compatible (actual => ACTUAL_Odd_parity_output,
                     expected => WAV_expect_Odd_parity_output))
        report "Error on Odd_parity_output output"
              severity WARNING;

        IF ( Compatible ( ACTUAL_Odd_parity_output,
                      WAV_expect_Odd_parity_output) ) THEN
              FAIL_SIGNAL <='L'; ELSE FAIL_SIGNAL <='1';
        END IF;
   END PROCESS;


 Monitor_Even_parity_output:
   PROCESS(ACTUAL_Even_parity_output,
                  WAV_expect_Even_parity_output)
   BEGIN
        assert(Compatible (actual => ACTUAL_Even_parity_output,
                    expected => WAV_expect_Even_parity_output))
        report "Error on Even_parity_output output"
              severity WARNING;

        IF ( Compatible ( ACTUAL_Even_parity_output,
                    WAV_expect_Even_parity_output) ) THEN
           FAIL_SIGNAL <='L'; ELSE FAIL_SIGNAL <='1';
        END IF;
   END PROCESS;

END parity_generator_checker_test;
```

Figure 6-23. The Testbench for the 54/74180 Parity Checker/Generator

6.4.3 Simulation Results

All of the WAVES files, the VHDL model, and the testbench are then compiled into the work library, and the simulation may then be carried out. The results of the simulation of the 8-bit parity generator/checker are shown in Figure 6-24. Through this example, notice that the necessary WAVES files to test and simulate this model have been written with minimal effort. All the source code files for this example are included on the companion CD-ROM in a directory called "***parity***".

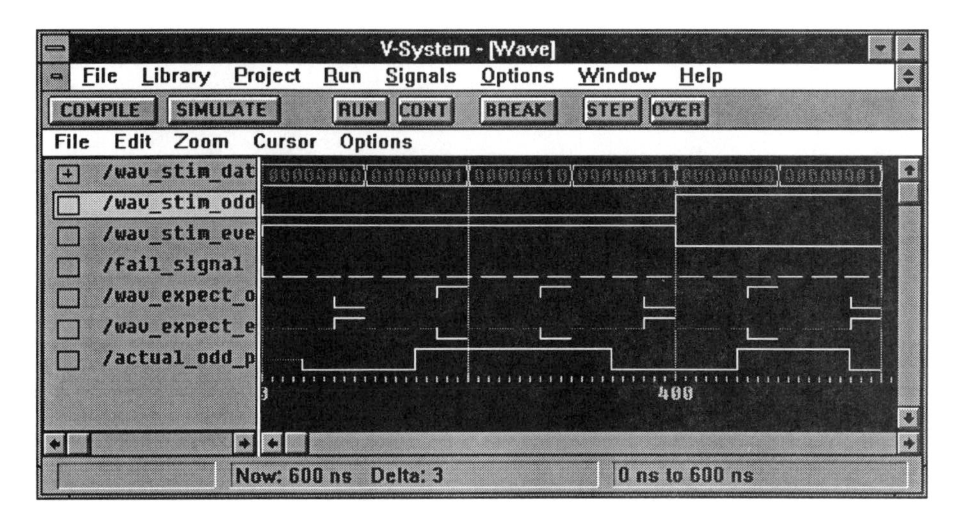

Figure 6-24. Simulation results for the 54/74180

In ***summary of our chapter on WAVES-VHDL integration and simulation***, we have covered all the essentials. We began with a conceptual view, explaining how all the elements in a WAVES-VHDL testbench interface during simulation. Next, we presented a group of interface functions necessary to implement the testbench in the WAVES-VHDL simulation environment. Then, we described an implementation of a WAVES-VHDL simulation system, demonstrating that the necessary WAVES files to test and simulate a design can be generated with minimal effort. Finally, we concluded the chapter with an additional example to further assist us in developing and executing the WAVES-VHDL simulation.

Throughout these discussions, and those of the earlier chapters, we have dealt with the external file in a cursory manner, basically implying that its purpose is to provide slice times associated with pin code sets (vectors). Next, in Chapter 7, we elaborate upon the external file, and explore its additional purposes and uses.

CHAPTER 7. THE EXTERNAL FILE

An aid to efficiency and utilization in WAVES

After completing most of the basic tutorial portion of the text, we have a good understanding of the WAVES dataset and testbench generation process, and how they support test and design environments. As we discussed in the previous chapter, we need a WAVES dataset and testbench to carry out simulation for a UUT (a VHDL model for the UUT). We presented details of the WAVES dataset in Chapter 5 and the testbench generation process in Chapter 6. Also, in Chapter 5, we provided illustrations of the three principal elements of the WAVES dataset: the WAVES files, the external files, and the header file. There, we described the WAVES files and header file in great detail, but only explained the basic functionality of the external file through an example. In this chapter, we provide a *complete view of the external file, to complete our basic tutorial.*

We devote a whole chapter to the external file because the complete depth and breadth of the external file constitute sufficient volume to warrant a chapter of their own. Basically, WAVES uses the external files to provide data, since external files are often more efficient at storing large amounts of data because there are fewer syntax restrictions. In fact, the only syntax restrictions are those imposed by the dataset itself. In this chapter we describe many legal shorthand notations to simplify generation of the external files. This notation will not only save time in generating the external files but will also conserve the resources needed to capture the external file. In particular, the size of the external file can be greatly reduced if these notations are utilized.

We begin our excursion through the external file with the *file declaration* required to use an external file in the WAVES dataset. Then, we introduce a formal description of the external file, as is described in Chapter 4 of IEEE STD 1029.1-1991 in the Backus Naur Form (BNF). Finally, we provide detailed explanations of the Level 1 external file format with plenty of examples. Using the information in this chapter, we will be able to create external files in the future with confidence and efficiency.

Before we proceed with the technical content of the external file, we present some supplemental information that we must keep in mind. First, the WAVES dataset does not *have to* contain any external files, although external files are indeed included in the majority of actual datasets. If the WAVES files contain all the information necessary to generate a waveform, an external file is not needed, but this is a rare instance. Next, if the external files *are* needed, the dataset must also supply a *reader procedure* that reads the external file and converts the data in the external file to standard WAVES objects. In this sense, the meaning of an external file is, therefore, dependent upon this reader. This dependency between the external file and the reader procedure is apparent in the D flip-flop example used in the basic tutorial portion of this book. Finally, the external file descriptions presented in this chapter are only applicable to WAVES Level 1. The aspects of WAVES Level 2 will be presented in Chapter 10. Now, let's begin with the formal description of the external file.

7.1 Required File Declarations

As we described in Section 5.3.2 of Chapter 5, we need to include a file declaration in the waveform generation procedure (WGP), to use an external file as a part of the WAVES dataset. There, we provided an example of the file declaration for the D-type flip-flop. The following is a more formal description of a file declaration. In this description, parameters or strings that we need to specify are indicated in italics.

```
file_declaration →
        FILE name : TEXT is in "name_string";
```

While the name string can be any string expression, it is recommended practice to use names that are portable across operating systems. For instance, short names consisting of alphanumerics are normally portable. The string could also be a (pointer) softlink (UNIX), which allows users to change external files to remove multiple test vector sets. The following file declaration shows the recommended practice.

```
        -- Specify the external test pattern file
        FILE vector_file : TEXT is in "vector.txt";
```

A file is open for reading when the procedure (WGP) in which it is declared starts. When the procedure exits, the file is closed. If the procedure is entered a second time then the file is opened again from the beginning.

In ***summary of our file declaration,*** the WAVES dataset must supply a procedure that reads the external file and converts the data in the external file to standard WAVES objects. Here, we described the file declaration required to allow the WGP to read an external file. Now, having established the necessary file declaration for the external files, we will discuss the external file format.

7.2 The External File Format

The WAVES Level 1 external file format was designed to support many of the common "tabular" or "truth table" waveform representations. These tabular or truth table waveform representations are frequently used when a test engineer develops a test program for Automatic Test Equipment (ATE). In this sense, WAVES level 1 was developed to support actual ATE environments. This ATE support feature becomes more obvious if we recall the concept of *slices* from the previous chapters. Each slice in the waveform can be viewed as a tester cycle in ATE environments. The external file format we explain in this chapter defines file slices that correspond to the waveform slices. However, it is very important to realize that the external file does not define slices directly, but rather defines intermediate objects called *file slices* that are manipulated by the waveform generator procedure in a user-specified manner.

As we discussed in Chapter 5, the meaning of the characters in an external file is determined by a method used to read data from the file. If the reader uses the WAVES procedure **Read_File_Slice**, which is defined in the package **Waves_Objects**, then the meaning of the characters in the file is as we describe in this section, and those characters define file slices. If the reader uses a different procedure (which is only possible in WAVES Level 2 as we illustrate in Chapter 10) then the same characters could have a different meaning. Therefore, the external file, by itself, has no meaning. It is important to remember that, in this section, we always assume that data from the file is being read by the **Read_File_Slice** procedure (WAVES Level 1). In this section, we concentrate upon the external file format that defines file slices. We begin with the formal description of the external file format which is described in Chapter 4 of IEEE STD 1029. We then provide many examples to illustrate efficient use of this external file format.

7.2.1 The Formal Description of the External File Format

The syntax which describes a single WAVES Level 1 external file format is given in Chapter 4 of IEEE STD 1029, and it is also provided below for the sake of completeness. The formal syntax is described in BNF. Now let's begin with the formal syntax. In this syntax, white space is permitted between any of the tokens. The syntax element *cr* denotes an end of line.

```
level_1_external_file →
        { <file_slice> <cr> }◊

file_slice →
        { <file_slice_command>};◊

file_slice_command →
        { <quoted_string> | <skip_command> | <timing_command> ◊

skip_command →
        = <integer> ◊

timing_command →
        : <integer> | : <time> ◊

quoted_string →
        " <character> [ <non_blank_string> ]
        | <non_special_character> <non_blank_string> ◊
```

A file slice may span multiple records, but every file slice must begin on a new record. Blank records in the file are also permitted. As shown above, a file slice can be built from three elements: **Quoted_Strings**, **Skip_Commands**, and **Timing_Commands**. Here, we provide a brief description of each element, deferring our discussions of their usage to the next section. We begin with the **Skip_Command**.

A **Skip_Command** is introduced with the "=" character and is followed by an integer. The integer indicates the number of the pin that will be loaded by the first character of the next quoted string. The valid integer range for the **Skip_Command** is between 1 (the first pin number) and the number of the last character position within a file slice. The integer must be followed by a space or end of record in order to distinguish it from subsequent commands.

The value of a **Timing_Command** is the value of the integer or time that follows the colon.

In general, the **Quoted_String** is used to specify a collection of pin codes that are to be applied to the test pins of a UUT within each slice. The value of a quoted string that begins with a double quote (") is the string formed from the character immediately following the quote and the following nonblank string. The value of a **Quoted_String** that does not begin with a double quote is the nonblank string itself. This implies that a space or end of record is required to delimit the right-hand side of the quoted string. Special characters such as the equal sign and the colon must be quoted to prevent the character from starting a skip or timing command. Other special characters that need to be quoted are (;), (%), and ("). A space can also be included in a pin code by prefixing it with a double quote.

Finally, comments are permitted in the external file. A comment begins with a comment operator and runs to the end of the record. The comment operator is a single percent character (%).

In ***summary of our formal description of the external file format,*** we have presented a rather cryptic, but complete, syntactical description of the external file. This formal description is necessary to unambiguously describe the external file format. Now, let's consider some examples that illustrate various ways to create external files.

7.2.2 External File Format Examples

Before we begin our examples, we need to understand the functionality and special features of the procedure **Read_File_Slice**. These functionality and features are essential in supporting efficient external file generation (such as data compression). As we described in Chapter 5, a file slice is loaded from an external file via a call to the **Read_File_Slice** procedure. This procedure reads commands from the file that cause it to ***update*** the named file slice. This is very different behavior from ***overwriting*** the named file slice because the file format permits fields to be omitted. If a field is omitted in the file, **Read_File_Slice** does not alter the corresponding field in the named file slice, thus preserving the previous value of that field. This update feature provides an effective form of data compression in the external file format. The simplest file slice consists only of a semicolon (;) which means "repeat last slice." Having discussed the update feature of the **Read_File_Slice**, we are now ready to explore some examples.

Our first example, shown in Figure 7-1, is for a dataset containing ten pins and defining the pin codes constant as the string "X01ZWLH-". This example generates eight slices. Each file slice begins on a new line and ends with a semicolon. The codes for the ten pins appear in order at the beginning of the file slice. The colon introduces the file slice timing (its duration). In this example, the first four vectors have a file slice duration of 20ns, the next two have durations of 25ns, and the last two revert back to the duration of 20ns. By now, we are very familiar with this form of the external file since we used it exclusively in the basic tutorial chapters. In Chapter 5, we also pointed out the importance of the space right before the colon, without providing any particular explanation or justification. Now we discover the reason we need that space. According to the definition of a **Quoted_String**, embedded special characters are treated as part of the string. Therefore, a space or end of record (cr) is required to delimit the right-hand side of the quoted string. Without the space, the colon for the timing command will be interpreted as a part of the quoted string (such as pin codes), which is not desirable.

```
0Z0000LLLL : 20 ns;
0Z0000HLLL : 20 ns;
0Z0001LLLH : 20 ns;
0Z0001HLLH : 20 ns;
0Z0010LLHL : 25 ns;
0Z0010HLHL : 25 ns;
0Z0011LLHH : 20 ns;
0Z0011HLHH : 20 ns;
          ↑
           This space is required to
          delimit the quoted string.
```

Figure 7-1. A Simple External File

Here, we realize that each vector is actually a change to the previous vector. This is a very important feature of the file format. It enables us to use the data compression capability which we delineated when we described the procedure **Read_File_Slice** in the beginning of this section. This means that only the changes need to be specified. Of course, the first vector must specify all information needed by the dataset, since there is no previous vector. Therefore, the example in Figure 7-1 can be written in a more succinct form as depicted in Figure 7-2.

```
0Z0000LLLL : 20 ns;
0Z0000HLLL ;
0Z0001LLLH ;
0Z0001HLLH ;
0Z0010LLHL : 25 ns;
0Z0010HLHL ;
0Z0011LLHH : 20 ns;
0Z0011HLHH ;
```

Figure 7-2. Specifying changes only in the External File

Another mechanism for compressing the external file has to do with skipping over repetitious portions of the vector string. Often some pin codes do not change from vector to vector. These pin codes can be omitted from the file slices by skipping over them. Because the file format is not fixed field (text position is irrelevant), a mechanism is needed to specify which pins are being skipped. This is accomplished by using the skip command. The skip command starts with an equal sign ('='). A skip integer follows the equal sign and the integer indicates the starting pin for the codes, after the skipped portion. Figure 7-3 illustrates the use of the skip commands for the same example we started in Figure 7-1. Here, "=6" indicates skipping (over the first 5 characters) to the 6th character, "=5" indicates skipping (over the first four characters) to the 5th character, and so forth.

```
0Z0000LLLL : 20 ns;
=6    0HLLL ;
=6    1LLLH ;
=6    1HLLH ;
=5   10LLHL : 25 ns;
=6    0HLHL ;
=6    1LLHH : 20 ns;
=6    1HLHH ;
```

Figure 7-3. Examples of Skip Commands in the External File

In this example, the second file slice starts loading codes at pin 6 (the first pin is numbered 1). Subsequent pin codes are copied into subsequent pin locations. Therefore, the second file slice loads code "0" for pin 6, the code "H" for pin 7, and the code "L" for pins 8 through 10. The result of this example is identical to that of the previous example. Spacing in this example was used to make pin code columns line up in order to improve readability of the pin codes. There is no need to line up columns, and in fact such spacing is wasteful since a space character takes up as much file storage as a pin code character. Hence, multiple spaces may always be reduced to a single space, so the example in Figure 7-3 can be written as we show in Figure 7-4.

```
0Z0000LLLL : 20 ns;
=6 0HLLL ;
=6 1LLLH ;
=6 1HLLH ;
=5 10LLHL : 25 ns;
=6 0HLHL ;
=6 1LLHH : 20 ns;
=6 1HLHH ;
```

Figure 7-4. Skip Commands without Columns Lined Up

Now, let's expand our example a little bit further. The intent of the following example is to illustrate the capability of various external file formats. A file slice can contain several instances of indexing, and the pin numbers specified do not have to be in any particular order. The following commands create the second file slice of the example shown in Figure 7-1, which was "`0Z0000HLLL : 20 ns;`" as originally configured.

```
=8 L =7 H =6 - =1 0Z000 =5 00 : 20 ns;
```

There are several interesting things to note in this example. First, note that the pin numbers were in no particular sequence. Also, note that the commands "`=6 -`" and "`=5 00`" both specified a pin code for pin 6. The first command specified the pin code "-" and the second specified the pin code "0". Commands within a file slice are always processed left to right and later commands are free to overwrite earlier commands.

Previously, we claimed that "white space" is wasteful. Sometimes, however, it is very useful and necessary to improve the readability of the test vectors. For instance, we may need to create test vectors for high pin-count devices such as microprocessors and application specific integrated circuits (ASICs). Often, such devices contain pins that can be logically grouped. For example, a microprocessor may have 32-bit address bus lines, 32-bit data bus lines, and 9 control signals. In this case, white spaces can be utilized to visually group these pins to improve readability. Also, note that vectors for each slice can occupy multiple lines. This is possible because a colon (:) is used to indicate a timing token and a semicolon (;) is used as file slice record terminator. Figure 7-5 illustrates an example utilizing white spaces to improve readability, and it also shows vectors occupying more than one line. In this example, we also utilize comments to improve readability.

```
% control_signal(9) address_bus(32-bits) data_bus(32-bits)
010111001 00000010 10111110 01111110 0000111 ZZZZZZZZ ZZZZZZZZ
       ZZZZZZZZ ZZZZZZZZ : 50 ns;
010111001 00000010 10111110 01111110 0000111 -------- 01010101
       00001111 11110000 : 60 ns;
```

Figure 7-5. Use of Space and Vectors Occupying Multiple Lines

Now, let us consider an example that involves special characters. As we described in the previous section, there are five special characters defined for use in external files. They are (:), (;), ("), (=), and (%). We have explained and used all these special characters, except the double quote ("). The double quote, as a part of a quoted string, provides an ***Escape*** mechanism for using any special character as a valid pin code in a vector. Figure 7-6 illustrates a vector which contains (=) as a valid pin code. In this example, pins 5 through 8 are assigned with a valid pin code (=). Here, the equal sign is not a skip command indicator.

```
1010 "==== 1110 X ZZ 10 : 35 ns;
```

Figure 7-6. A Special Character(=) as a Valid Pin Code

As usual, we can utilize the skip command (=) to define the next vectors, which provides data compression capability. Figure 7-7 illustrates this data compression capability and Figure 7-8 shows equivalent vectors without data compression (i.e., without use of the skip command).

```
1010 "==== 1110 X ZZ 10 : 35 ns;      ( = is pincode)
=9 0001                               ( = is skip command)
```

Figure 7-7. Data Compression using Skip Command (=)

```
1010 "==== 1110 X ZZ 10 : 35 ns;
1010 "==== 0001 X ZZ 10 : 35 ns;
```

Figure 7-8. Equivalent Vectors without Data Compression

So far, all the examples we presented utilized the timing command with an actual time indicated (20 ns, 35 ns, etc.). However, we note from the formal syntax, the timing command can also use an integer. Time indicates the duration of the slice and the integer indicates a timing set selection. In general, multiple timing sets are defined in the waveform generator procedure (WGP). These multiple timing sets are necessary to define complex and realistic waveforms more efficiently. In realistic design and testing environments, we often face a situation that requires more than one timing set to describe the waveform. For example, we may have to deal with devices with bi-directional pins. A good example device would be a bi-directional transceiver or a pre-loadable shift-register. For this type of device, we need to have multiple timing sets to describe different modes of operation. Certain pins may act as input pins in one operation mode and as output pins in the other modes. In Figure 7-9, we provide the declaration of a timing set which consists of two frame sets for an arbitrary device. Here, we need not worry about the detailed syntax of this timing set declaration, but simply realize that there are two timing sets to choose from for each slice defined in the external file. A complete example utilizing multiple timing sets will be presented in Chapter 8. For now, let us concentrate upon the ways in which we can select appropriate timing sets via the external files.

```
Variable WTL : Wave_timing_list (1 to 2) := (
        --
   -- Frame Set array or timing set 1 for shift and
   -- hold operations
   -- The bi-directional pins acts as output pins in these modes
        --
        ( Delay  => Delay ( 100 ns ),
          Timing => New_Time_Data(
        New_frame_set_array(Pulse_high( 50 ns, 80 ns), Clock) +
        New_frame_set_array(Window( 85 ns, 95 ns), IO) +
        New_frame_set_array(Non_return( 5 ns), inputs) +
        New_frame_set_array(window( 85 ns, 95 ns), outputs)
        )),
        --
        -- Frame Set array or timing set 2 for load operation
        -- The bi-directional pins acts as input pins
        -- in this mode
```

```
--
(  Delay   => Delay ( 100 ns ),
   Timing => New_Time_Data(
 New_frame_set_array(Pulse_high( 50 ns, 80 ns), Clock) +
    New_frame_set_array(Non_return( 5 ns), IO) +
    New_frame_set_array(Non_return( 5 ns), inputs) +
    New_frame_set_array(window( 85 ns, 95 ns), outputs)
 )));
```

Figure 7-9. Declaration of timing sets with two frame set arrays

Let's begin an example, shown in Figure 7-10, which supports multiple timing sets. In this example, the first four vectors use index 1 and the next two use index 2, and the last two revert back to index 1. These index numbers are used to select a particular timing set from multiple timing sets declared in the waveform generator procedure. The first file slice states that the WGP shall use the first timing set and pin codes "0Z0000LLLL" in order to generate a slice of waveform.

```
0Z0000LLLL : 1;
0Z0000HLLL : 1;
0Z0001LLLH : 1;
0Z0001HLLH : 1;
0Z0010LLHL : 2;
0Z0010HLHL : 2;
0Z0011LLHH : 1;
0Z0011HLHH : 1;
          ↑
           This space is required to
             delimit the quoted string.
```

Figure 7-10. Simple External File showing Multiple Timing Sets

All the techniques presented in our previous examples, including the skip command and quoted strings, still apply here. For instance, Figure 7-11 illustrates an example of the data compression utilizing the skip command.

```
0Z0000LLLL : 1;
=5 0HLLL ;
=5 1LLLH ;
=5 1HLLH ;
=4 10LLHL : 2;
=5 0HLHL ;
=5 1LLHH : 1;
=5 1HLHH ;
```

Figure 7-11. Data Compression utilizing Skip Commands

In ***summary of our external file format examples***, we have presented useful illustrations which outline the efficient use of external files. We discussed three elements that make up file slices, with all the possible variations. These three elements are the quoted string, the skip command, and the timing command. In addition, we described the purpose and use of comments in the external file.

In ***summary of the external file format***, we presented a formal syntactical description of the external file format in this section. Then, we explained this formal description of the external file format with many examples to enhance our understanding of the external file. Now, we are able to create an external file efficiently.

In ***summary of this chapter***, we have described the complete depth and breadth of the external file. WAVES permits the use of external files to provide data. External files are often more efficient at storing large amounts of data because there are fewer syntax restrictions. Having gained a good understanding of the external files, we are now ready to generate the WAVES dataset for more complicated and realistic device testing. In Chapter 8, we describe relative edge placement within a slice, and illustrate WAVES' capability in handling bi-directional pins of devices.

CHAPTER 8. SOME PRACTICAL ISSUES AND EXAMPLES IN WAVES

Working with diverse, realistic timing situations

At this point in our text, we have completed the basic tutorial portion, and have described and demonstrated the fundamental WAVES and VHDL construction and execution processes. While the examples we presented in the basic tutorial portion of our text served well to illustrate the basic concepts of WAVES and its interaction with VHDL models (i.e., UUT), we did not explain WAVES' capability to handle more realistic and complex timing issues. Essentially, we did not wish to add complexity to these examples which would have distracted from understanding the basic WAVES concepts we were demonstrating. In Chapter 7, we also described the complete anatomy of the external files, to assist in creating efficient test vectors. Now, we are armed with knowledge and ready to add some realism.

In this chapter we introduce the issues relating to more realistic, practical, and complex timing relationships which WAVES is able to handle. We begin with an example illustrating relative edge placement within a slice. Frequently, an edge transition of a signal in a waveform is specified with respect to an edge transition of another signal. WAVES provides a mechanism to describe this type of signal transition *via* the relative edge placement. We demonstrate this feature in our first example. Then, we present another example illustrating WAVES' capability to handle the bi-directional nature of a device. More often than not, a reasonable design will contain one or more bi-directional devices such as microprocessors, transceivers, and shift registers. Consequently, it is important to learn how to support these bi-directional devices using WAVES. Now, let's begin with the first example, relative edge placement.

8.1 Relative Edge Placement in WAVES

All the examples we presented in the basic tutorial portion of our text provided a simplified way to specify timing and edge transitions in the waveform. More specifically, the timing and edge transitions were specified with respect to the beginning of each slice. (Recall the frame set array declaration within the waveform generator procedure of each example.) This method works fine and efficiently if a waveform we want to describe is simple, or if the waveform can be sliced such that all timing and edge transitions can be specified with respect to the beginnings of the slices. Sometimes, however, it is not practical to vary the lengths or duration of the slices. Furthermore, most design specifications (such as data books) of reasonably complex devices specify the timing and edge transitions of signals with respect to one or the other, as shown in Figure 8-1. (tp is defined with respect to the falling edge of the signal 1.) This means, at the very least, we have to translate these relative timing and edge transitions into ones with respect to the beginnings of the slices, in order to use the methods we described in the basic tutorial. Figure 8-2 illustrates this translation process using the same waveform shown in Figure 8-1. (Here, we assumed t is 10 ns.) We see that this manual translation process can be time consuming and error-prone. The good news is that WAVES provides a mechanism to handle these relative timing and edge transitions as-is. We call this mechanism the "***relative edge placement***" technique, which we now explain with an embedded microprocessor example.

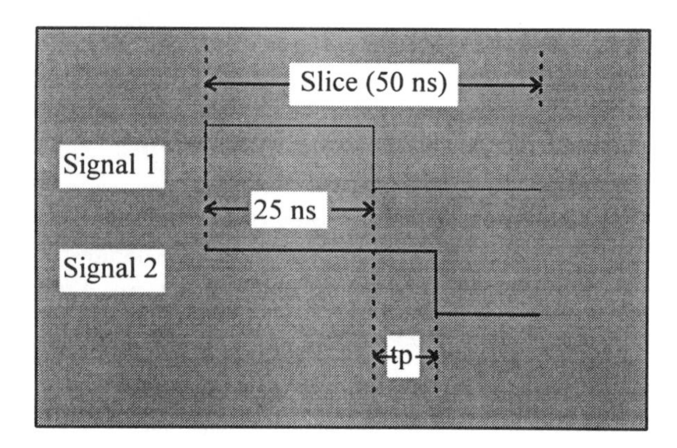

Figure 8-1. Relative Timing and Edge Transition

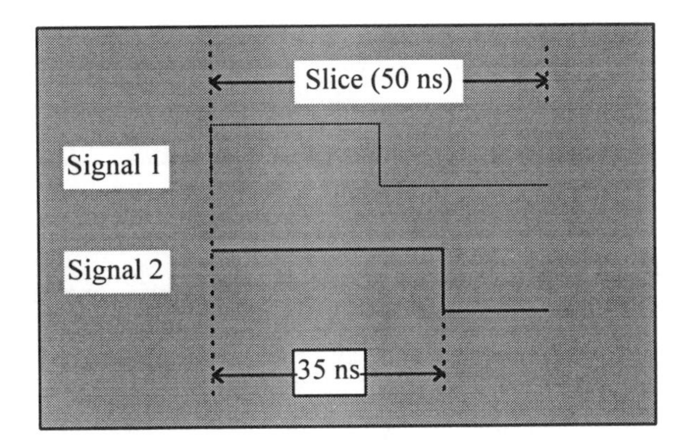

Figure 8-2. Translated Timing and Edge Transition

Here, we again follow the same three distinct logical design steps we described in Section 6.3 of Chapter 6. These steps are: (1) developing VHDL models for the UUT, (2) generating the WAVES dataset, and (3) generating the testbench. First, we present the device specifications which describe the functionality and input/output interfaces. Then, we develop all the necessary WAVES and VHDL files to support simulation of the WAVES-VHDL design. All the source code files for this example are included on the companion CD-ROM in a directory called "***Micro_P***".

8.1.1 Device Specifications of an Embedded Microprocessor

In order to demonstrate WAVES' capability to handle the relative timing and edge transitions among signals, we hand-crafted an embedded microprocessor example. Of course, designing a complete embedded microprocessor system is beyond the scope of this text. In fact, thousands of highly-trained and experienced professionals may devote their entire careers to developing embedded systems. Here, our intent is to develop a simple example and exercise WAVES to describe relative timing and edge transitions. Therefore, our example covers only a fraction of typical microprocessor functionality, although we can add additional functionality to this example if we choose to use it elsewhere. Here, we do not wish to add complexity to this example which may distract from understanding the WAVES concepts we want to demonstrate. Now, let's begin describing the actual example.

Our embedded microprocessor is a 21-pin device with 8 data bus inputs (**In_Port_0** through **In_Port_7**), 2 control inputs (**CLK** and **DACK**), 8 address bus outputs (**Addr_Bus_0** through **Addr_Bus_7**), and 3 outputs (**R_W**, **AS**, and

Read_Finish). Figure 8-3 lists the name of the signals available on the embedded microprocessor, and provides a short description of their functionality.

Pin Names	Description
CLK	This is an input **clock** signal that drives all internal timing.
DACK	**Data Acknowledgment** signal. This input signal is used to notify the processor that the data on the In_port is valid.
In_port (7 downto 0)	This is 8-bit wide **input bus signal** set from which the processor reads data.
R_W	The processor sends a high level ('1') on this signal when it wants to **read**, otherwise this signal is floated ('Z').
AS	The processor sends a low level ('0') on this signal to indicate the **address** on the address bus is **valid**, otherwise this signal is floated ('1').
Addr_Bus (7 downto 0)	The processor sends out an **8-bit wide address** on these signals.
Read_Finish	This output signal is used to indicate when the **read operation is finished**. (Low level when the read operation begins and high level when the read operation is finished.)

Figure 8-3. Signal Names and Description of the Embedded Processor.

In our example, this processor is dedicated to monitor a network device or peripheral sitting on a specific address space. Figure 8-4 illustrates the system interfaces between the processor and other components in the system. For every clock cycle, the processor reads the status of the network device through its **In_Port** (8-bit wide) and notifies the host system when it finishes reading the network device *via* the signal **Read_Finish**. Now, we provide a brief description of the read operation of the processor where the relative timing and edge transitions are discussed.

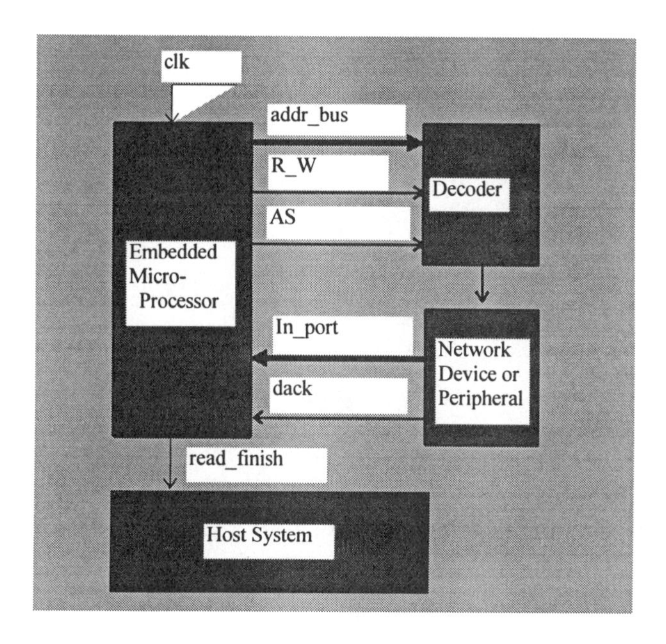

Figure 8-4. System Interfaces between the Processor and other Components

The processor initiates the read operation (or *cycle*) by sending out a high-level signal on the **R_W** port. Next, it sends out the address of the network device from which it wants to read *via* the address bus (**Addr_Bus_0** through **7**). Then, the processor sends out a low-level signal on the AS. These three signals (**R_W**, **AS**, and **Addr_Bus**) are used by the decoder to select the network device. Then, the processor waits until the signal **DACK** becomes active (a low level on the **DACK**). Here, the processor is waiting until the data on the **In_Port** are valid. While the processor is waiting, the selected network device sends out status data on the **In_Port** and it enables the signal **DACK** to notify the processor that the data on the **In_Port** is valid. Next, the processor "wakes up" and reads data from the **In_Port**. Then, the processor disables all the signals related to the read operation until a new read operation is initiated. Finally, it uses the signal **Read_Finish** to notify the host system that it has finished the read operation.

So far, we have described a sequence of events on multiple signals, involved in the read operation, without considering exact timing and edge transitions among these events. Here, we specify the timing and edge transitions among these events with respect to one another. Figure 8-5 illustrates the timing diagram of the read operation in detail. Please note that the timing of one signal is specified with respect to the edge transition of another signal. Figure 8-6 lists the exact timing information for each edge transition. This information will be used to develop the VHDL model and the WAVES dataset for this embedded processor example.

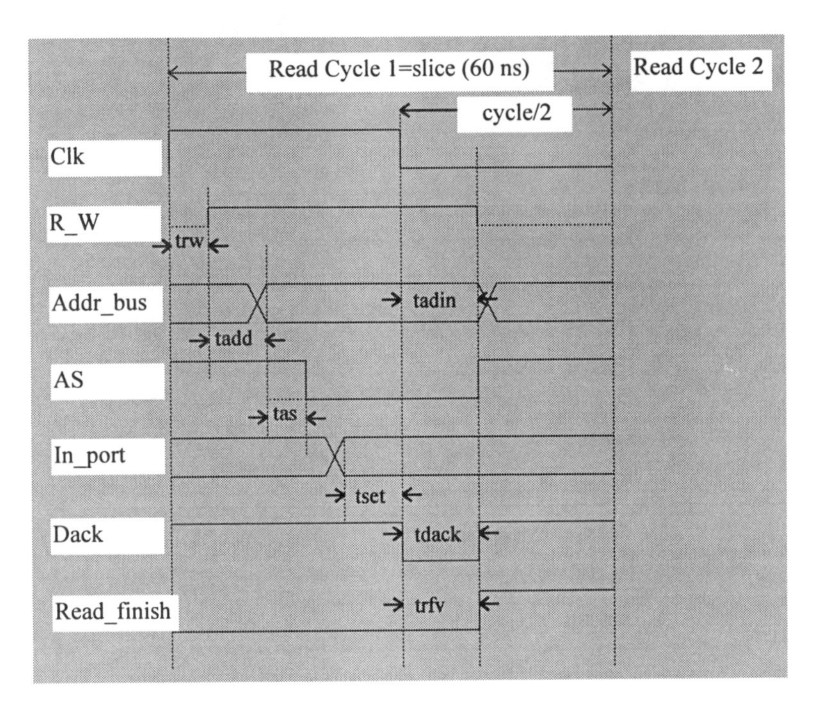

Figure 8-5. Timing diagram of the Read Cycle

Description	Symbol	Time (ns)
Read Enable	trw	3
Address Enable	tadd	4
Address Strobe Enable	tas	5
Data Setup	tset	5
Address Invalid	tadin	10
Data Acknowledgment	tdack	10
Read Finish Valid	trfv	10
Clock Cycle	cycle	60
Half Clock Cycle	cycle/2	30

Figure 8-6. Timing information summary for the Read Cycle

In ***summary of our device specifications***, we presented the functionality, timing and edge transition information, and input/output interfaces for the embedded processor through the device specifications. In particular, we used the relative timing for the timing and edge transition specifications. Next, we develop a VHDL model (a UUT) based on the device specifications presented above. Following the VHDL model development, we will develop the WAVES dataset and testbench to test the UUT.

8.1.2 VHDL Model Generation for the Embedded Processor

As we described in Chapter 6, the first step toward the simulation of a WAVES-VHDL design is to develop a VHDL model for the UUT. We first developed the behavioral description of the embedded microprocessor based on the specifications described in a previous section, and it appears in Figure 8-7. We also added additional comments to capture useful information such as title, file name, date, author, organization, and so forth. These comments are not required; however, we have provided them for adequate documentation.

```
-- Device :          Embedded Processor Monitoring a Network Device
-- File Name:        micro_p.vhd
-- DATE:             Fri Mar  8 09:25:58 1996
-- Organization:     Company XYZ
-- AUTHOR            Design Automation Group

Library ieee;
use ieee.std_logic_1164.all;

entity micro_p is
  port (CLK : in std_logic;
       DACK : in std_logic;
       In_port : in std_logic_vector (7 downto 0);
       R_W: out std_logic;
       AS: out std_logic;
       ADDR_BUS: out std_logic_vector (7 downto 0);
       READ_FINISH: out std_logic);
end micro_p;

architecture behavior of micro_p is
begin
  process(clk, dack)
    constant Trw: time := 3 ns;
    constant Tad: time := 4 ns;
    constant Tas: time := 5 ns;
    constant Disable: time:= 10 ns;
    variable DATA_REGISTER: std_logic_vector (7 downto 0);
```

```
     begin
      -- Initiate READ
       if ((clk = '1') AND (clk'event)) then
          READ_FINISH <= '0';
          R_W <= '1' after Trw;
          ADDR_BUS <= "01010101" after Tad + Trw;
          AS <= '0' after Trw + Tad + Tas;
       end if;

      --  Read Data
       if ((dack = '0') AND (dack'event)) then
          DATA_REGISTER := In_port;

      -- Add any data manipulation or checking
            --   routine here

       end if;

      -- Finish READ
       if ((clk = '0') AND (clk'event)) then
          R_W <= 'Z' after Disable;
          ADDR_BUS <= "ZZZZZZZZ" after Disable;
          AS <= '1' after Disable;
          READ_FINISH <= '1' after Disable;
       end if;

end process;
end behavior;
```

Figure 8-7. Behavioral Model for the Embedded Processor

In this example, we utilize a process (a VHDL construct) to describe the behavior of the embedded processor. The first part of the process (i.e., the first if statement) waits for a rising edge of the clock then the processor sends out appropriate signals (**R_W**, **Addr_Bus**, and **AS**) with specified timing, to initiate the read operation. Note that the processor always sends out a fixed address of "01010101". This is due to the fact that the processor is dedicated to monitor the network device and the fixed address ("01010101") is assigned to the network device. The second part of the process (the second if statement) waits for a falling edge of **DACK** (data acknowledgment), then the processor reads data transmitted by the network device from the **In_Port**. (When the read operation is initiated by the processor, the network device transmits data to the processor via **In_Port**, then it sends the signal **DACK** to notify the processor that data is "good" to be read.) Utilizing this data, the processor may carry out additional checking or manipulation to meet application needs. This additional checking or manipulation is not shown here because it is an application-specific issue and beyond the scope of this text. The third part of the process waits for a falling edge of the clock, then the processor sends out the appropriate signals with associated delay to finish the read operation. The actual VHDL code of this processor is included on the companion CD-ROM with file name "**Micro_P.Vhd**" in the ***Micro_P*** directory.

In ***summary of VHDL model generation***, we have presented a simple embedded microprocessor VHDL model which was created based on the device specifications. We also emphasized that the model developers should always provide adequate comments to capture useful information about the VHDL model and to make the model readable. Now, let us generate the WAVES dataset, which is the second logical step, to support the WAVES-VHDL design simulation.

8.1.3 WAVES Dataset Generation

After we develop a VHDL model (the UUT), the next step is to generate a corresponding WAVES dataset for the UUT. This is *the second principal step*, to support the WAVES-VHDL simulation. As we described in Chapter 5, the WAVES dataset development procedure, which utilizes the **WAVES_1164** library, consists of five sub-tasks: creating a header file for documentation, creating and compiling the test pins package, modifying and compiling the WAVES Objects package, creating and compiling the waveform generator package, and creating an external (test vector) file. Here, we present each WAVES dataset element generated by each sub-task for our example. We begin with the header file.

8.1.3.1 Header File for the Embedded Microprocessor Example

As usual, the header file captures information that is not available within WAVES files or the external files, yet that is necessary or useful to describe the WAVES dataset completely. Such information includes the dataset identification and construction information, external test vector file identification, and waveform generator procedure identification. We illustrate the header file for the embedded microprocessor example in Figure 8-8.

```
-- ***********************************************
-- ******** Header File for Entity: micro_p
-- ***********************************************
--
-- Data Set Identification Information
--
TITLE         A General Description
DEVICE_ID     micro_p

DATE          Fri Mar  8 09:25:58 1996
ORIGIN        Company X Design Team
AUTHOR        Company or Person
AUTHOR        Maybe Multiple ... Companies or People
DATE          Fri Mar  8 09:25:58 1996
ORIGIN        Modified by Company X Design Team
AUTHOR        Who did it Company or Person

OTHER         Any general comments you want
OTHER         Built Using the WAVES-VHDL 1164 STD Libraries
--
-- Data Set Construction Information
--
WAVES_FILENAME /micro_p_pins.vhd WORK
library          WAVES_1164;
use              WAVES_1164.WAVES_1164_Pin_Codes.all;
use              WAVES_1164.WAVES_1164_Logic_Value.all;
use              WAVES_1164.WAVES_Interface.all;
use              WORK.UUT_Test_pins.all;
WAVES_UNIT       WAVES_OBJECTS   WORK
WAVES_FILENAME /micro_p_wgen.vhd WORK
--
EXTERNAL_FILENAME vector.txt                   VECTORS
--
WAVEFORM_GENERATOR_PROCEDURE     WORK.waves_micro_p.waveform
```

Figure 8-8. The Header File for the Embedded Microprocessor

8.1.3.2 Test Pins Package for the Embedded Microprocessor Example

The **Test_Pins** type is the first source code of the WAVES dataset that we need to define and compile. It defines all the input and output pins of the unit under test, and provides a communications interface between the UUT, the WAVES dataset, and the testbench for the waveform information. The stimulus waveform information is applied to the UUT through the input pins, and the expected response waveform information is observed upon the output pins. More precisely, the waveform information of the UUT is applied and observed through the WAVES port list and the **Test_Pins** type is used to define the WAVES port list. We show the **Test_Pins** declaration for the embedded microprocessor example in Figure 8-9. Note that we split all the bus signals into individual signals within the **Test_Pins** type. According to the port declaration of the **Micro_P** entity previously given in Figure 8-7, the **In_Port** and **Addr_Bus** are 8-bit wide bus signals. These bus signals are represented as a collection of individual signals for each bit slice in the **Test_Pins** type. The individual pin representation is a requirement imposed in WAVES Level 1 to support Automatic Test Equipment.

```
-- ******** This File Was Automatically Generated  ********
-- ******** By The WAVES-VHDL Tool Set       ********
-- ******** Generated for Entity: micro_p
-- ******** This File Was Generated on: Fri Mar  8 09:25:58 1996
--
--
PACKAGE uut_test_pins IS
TYPE test_pins IS (CLK, DACK, In_port_7, In_port_6, In_port_5,
              In_port_4, In_port_3, In_port_2, In_port_1,
              In_port_0, R_W, AS, ADDR_BUS_7, ADDR_BUS_6,
              ADDR_BUS_5, ADDR_BUS_4, ADDR_BUS_3, ADDR_BUS_2,
              ADDR_BUS_1, ADDR_BUS_0, READ_FINISH);
END uut_test_pins;
```

Figure 8-9. The Test Pins Package for the Embedded Microprocessor

8.1.3.3 Waves_Objects Package for the Embedded Microprocessor Example

The second step in building the WAVES dataset is to add the proper context clauses to the beginning of the **Waves_Objects** package and compile this package into the work library. The context clauses provide visibility to the **Waves_Interface**, **Logic_Value**, **Pin_Codes**, **Test_Pins** packages. Appropriate context clauses for the embedded microprocessor example are given in Figure 8-10. These are, of course, not necessary if we are using the 1164 packages.

```
use STD.TEXTIO.all;
library WAVES_STD;
use WAVES_STD.WAVES_SYSTEM;

-- A context clause providing visibility to an analyzed copy of
-- WAVES_INTERFACE is required at this point.
-- Context clauses providing visibility to LOGIC VALUE,
-- TEST PINS,
-- and PIN CODES are required at this point.

Library WAVES_1164;
use WAVES_1164.WAVES_Interface.all;
use WAVES_1164.WAVES_1164_Logic_Value.all;
use WAVES_1164.WAVES_1164_Pin_Codes.all;
use WORK.UUT_Test_pins.all;
```

Figure 8-10. The Context Clauses for the Waves_Objects Package

8.1.3.4 Waveform Generator Procedure (WGP) for the Embedded Microprocessor Example

The next step in building the WAVES dataset is to define and compile the waveform generator procedure. The waveform generator procedure reads a slice of pin codes from an external test vector file. Then, it constructs a slice of the waveform utilizing the definition of the **Frame_Set_Array** declarations, the outputs stimulus, and the expected responses to the testbench through the WAVES port list. We show the WGP for the embedded microprocessor example in Figure 8-11.

```
LIBRARY WAVES_STD;
USE WAVES_STD.WAVES_Standard.all;

USE STD.textio.all;
LIBRARY WAVES_1164;
USE WAVES_1164.waves_1164_frames.all;
USE WAVES_1164.waves_1164_pin_codes.all;
USE WAVES_1164.waves_interface.all;
USE work.waves_objects.all;
USE work.uut_test_Pins.all;

PACKAGE WGP_micro_p is
     PROCEDURE  waveform(SIGNAL WPL : inout WAVES_PORT_LIST);
END WGP_micro_p;
```

```
-------------------------------------------------------
PACKAGE BODY WGP_micro_p is

-- This is the uut pin declaration pin and ordering
-- Remember you need to match the External file to This order
--
--CLK, DACK, In_port_7, In_port_6, In_port_5, In_port_4,
--In_port_3, In_port_2, In_port_1, In_port_0, R_W, AS,
--ADDR_BUS_7, ADDR_BUS_6, ADDR_BUS_5, ADDR_BUS_4, ADDR_BUS_3,
--ADDR_BUS_2, ADDR_BUS_1,  ADDR_BUS_0, READ_FINISH

    PROCEDURE  waveform(SIGNAL WPL : inout WAVES_PORT_LIST) is

       FILE vector_file : TEXT is in "vector.txt";

       VARIABLE vector : FILE_SLICE := NEW_FILE_SLICE;

      -- declare time constants to use or use time literals
      -- constants or time literals can be used as
      -- the frame time values
      -- Constants used to address the relative timing

        CONSTANT cycle : time := 60 ns;
        CONSTANT half_cycle : time := cycle/2;
        CONSTANT data_setup : time := half_cycle - 5 ns;
        CONSTANT tdack : time := half_cycle + 10 ns;
        CONSTANT trw : time := 3 ns;
        CONSTANT tadd : time := 4 ns;
        CONSTANT tas : time := 5 ns;
        CONSTANT tfinish : time := half_cycle + 10 ns;
        CONSTANT trfvalid : time := tfinish + 10 ns;

       -- Group the pins using pinset

   CONSTANT In_port: pinset:= new_pinset(( In_port_7, In_port_6,
                     In_port_5, In_port_4, In_port_3, In_port_2,
                     In_port_1, In_port_0));

        CONSTANT ADDR_BUS: pinset:= new_pinset(( ADDR_BUS_7,
          ADDR_BUS_6, ADDR_BUS_5, ADDR_BUS_4, ADDR_BUS_3,
                        ADDR_BUS_2, ADDR_BUS_1, ADDR_BUS_0));

       -- Define the frame set array utilizing the time
       -- constants
       -- Expressions formed with the time constants are used
       -- to handle the relative timing and edge transitions
```

```
  CONSTANT vector_FSA : Frame_set_array :=
    New_frame_set_array(Pulse_low(half_cycle, cycle), clk) +
    New_frame_set_array(Pulse_low(half_cycle, tdack), DACK) +
    New_frame_set_array(Non_return(data_setup), in_port) +
    New_frame_set_array(window(trw, tfinish), R_W) +
    New_frame_set_array(window(trw + tadd, tfinish), ADDR_BUS) +
    New_frame_set_array(window(trw + tadd + tas, tfinish), AS) +
    New_frame_set_array(window(tfinish, trfvalid), READ_FINISH);

      VARIABLE timing : time_data := new_time_data(vector_fsa);

     BEGIN
       loop
         -- get first vector
         READ_FILE_SLICE (vector_file, Vector);
         exit when vector.end_of_file;
         apply(wpl, vector.codes.all,
                    Delay(vector.fs_time), timing);
        end loop;

     END waveform;

END WGP_micro_p;
```

Figure 8-11. Waveform Generator Procedure for the Embedded Microprocessor

As mentioned before, the WGP requires visibility of the **Waves_Standard**, **Textio**, **Waves_1164_Frames**, **Waves_1164_Pin_Codes**, **Waves_Interface**, **Waves_Objects**, and **UUT_Test_Pins** packages. The visibility is ensured by adding appropriate context clauses to the beginning of the WGP package.

In this example, we declare the external test vector file (vector.txt) that the WGP accesses for pin codes. Next, we declare time constants (cycle, **Half_Cycle**, **Data_Setup**, tdack, trw, tadd, tas, tfinish, trfvalid) to efficiently handle the relative timing and edge transitions among signals. We recall from the timing diagram of the read cycle in Figure 8-5, that many timing parameters were defined with respect to an event on another signal. For example, we defined the address enable time (tadd) from the rising edge of the signal **R_W** to the time that the address on the **Addr_Bus** is valid. From this example, it is clear that WAVES should be able to accommodate these relative timing issues. WAVES handles these relative timing issues by using the time constants. These time constants are used to define the frame time values within each frame set. Soon, we will demonstrate the use of the time constants clearly when we define the frame set array.

Next, we group the input and output pins based on the timing characteristics of each signal (i.e., define the pinsets). This grouping of the pins enables us to use the same frame set (i.e., waveform shape) for pins with similar timing and electrical characteristics. In this example, it makes sense to group all **In_Port** (0 to 7) and **Addr_Bus** (0 to 7), respectively, since they have similar timing characteristics.

Then, we declare a frame set array constant, **Vector_FSA**, which associates test pins and frame set. We also assign the edge transition time to each frame set to uniquely define the frame set. As we mentioned above, we utilize the time constants to define the frame time values (the edge transitions time) within each frame set, which enable us to support the relative timing and edge transition efficiently. For instance, many of the timing parameters are defined with respect to the half cycle of the clock (such as **Data_Setup** time, trfv...). We used the time constant **Half_Cycle** to define these timing parameters. We can also form an expression utilizing the time constants and use this expression to define the frame time value for the frame set. Here is an example using an expression for defining the frame time.

```
New_frame_set_array(window(trw + tadd + tas, tfinish), AS)
```

In this example, the edge transitions (t1,t2) of the frame set **window** for the signal AS are defined using expressions formed with time constants. In particular, t1 is defined with an expression “trw + tadd + tas”. This is a very important and useful feature in describing the relative timing and edge transitions which are common in complex devices. Using this type of expression, constructed with the time constants, we can avoid the tedious and cumbersome translation process described earlier in Figure 8-2.

Finally, the WGP reads the external file one slice of pin codes at a time, and constructs each slice of waveform utilizing the APPLY function. The constructed waveform is provided to the testbench to drive the WAVES-VHDL simulation. This process repeats until the end of the external file is reached.

8.1.3.5 External Test Vector File for the Embedded Microprocessor Example

The final step in building the WAVES dataset is to create an external test vector file which contains the pin codes. When we create the external file, we must remember to match the pin codes to the order of the test pins declaration. The external file used for testing the VHDL model of the embedded microprocessor is shown in Figure 8-12. Each slice in this file represents stimulus and expected data for a read cycle. For example, the first slice tests the read operation of the processor by providing appropriate data on the CLK, DACK, and **In_Port** pins. At the same time, it also provides the expected data for the output pins (**R_W**, **AS**, **Addr_Bus**, and **Read_Finish**). Then, the testbench compares the expected data with the actual outputs to verify the operation of the embedded microprocessor.

```
%CLK DACK IN_PORT R_W AS ADDR_BUS(8) READ_FINISH
0 0 00001111 1 0 01010101 1 : 60 ns;
0 0 11110000 1 0 01010101 1 : 60 ns;
0 0 00110011 1 0 01010101 1 : 60 ns;
0 0 00111100 1 0 01010101 1 : 60 ns;

% Add more vectors  .......
```

Figure 8-12. External Test Vector file for the Embedded Microprocessor

In ***summary of WAVES dataset generation***, we can see that a systematic approach we described in Chapter 5 is applicable in creating the WAVES dataset regardless of the complexity of the device. We have also presented the capability and flexibility of the WAVES dataset in handling more complex timing issues, such as the relative edge placement. The relative edge placement technique utilizes time constants and the expressions formed with the time constants. Next, we create a testbench that utilizes both the VHDL model of UUT and the WAVES dataset to carry out the simulation.

8.1.4 Testbench Generation

The *third and final principal step* required to simulate the WAVES-VHDL design is to generate a testbench. The testbench establishes communication channels between the WAVES dataset and the UUT VHDL model. It also monitors the response of the UUT and compares it with the expected response from the WAVES dataset. In this example, our testbench consists of seven elements: context clauses, UUT component and configuration declarations, connection signal declarations, invocation of the waveform generator, translation functions, the UUT component instantiation, and monitor processes. Each element was described in detail in Section 6.3 of Chapter 6. Consequently, we simply present the testbench for the embedded microprocessor in Figure 8-13, without any further explanation. If required, we can find explanations of the various elements in Chapter 6.

```
LIBRARY ieee;
USE ieee.std_logic_1164.ALL;

LIBRARY waves_1164;
USE waves_1164.WAVES_1164_utilities.all;

USE WORK.UUT_test_pins.all;
USE work.waves_objects.all;
USE work.WGP_micro_p.all;

ENTITY test_bench IS
END test_bench;


ARCHITECTURE micro_p_test OF test_bench IS

-- ******************************************
-- ********** Configuration ****************
-- ******************************************

COMPONENT micro_p
  port (CLK : in std_logic;
       DACK : in std_logic;
       In_port : in std_logic_vector (7 downto 0);
       R_W: out std_logic;
       AS: out std_logic;
       ADDR_BUS: out std_logic_vector (7 downto 0);
       READ_FINISH: out std_logic);
end COMPONENT;

 -- Modify entity use statement

FOR ALL:micro_p USE ENTITY work.micro_p(behavior);

  --*********************************************************
  -- stimulus signals for the waveforms mapped into UUT INPUTS
  --*********************************************************

    SIGNAL WAV_STIM_CLK           :std_logic;
    SIGNAL WAV_STIM_DACK          :std_logic;
    SIGNAL WAV_STIM_IN_PORT      :std_logic_vector(7 downto 0);

  --*****************************************************
  -- Expected signals used in monitoring the UUT OUTPUTS
  --*****************************************************

    SIGNAL FAIL_SIGNAL                  : std_logic;
```

```
    SIGNAL WAV_EXPECT_ADDR_BUS  :std_ulogic_vector(7 downto  0);
    SIGNAL WAV_EXPECT_R_W              :std_logic;
    SIGNAL WAV_EXPECT_AS               :std_logic;
    SIGNAL WAV_EXPECT_READ_FINISH      :std_logic;


  --*****************************************************
  -- UUT Output signals used In Monitoring ACTUAL Values
  --*****************************************************
    SIGNAL ACTUAL_ADDR_BUS        :std_logic_vector(7 downto 0);
    SIGNAL ACTUAL_R_W                 :std_logic;
    SIGNAL ACTUAL_AS                  :std_logic;
    SIGNAL ACTUAL_READ_FINISH         :std_logic;
  --*********************************************************
  -- WAVES signals OUTPUTing each slice of the waves port list
  --*********************************************************

        SIGNAL wpl  : WAVES_port_list;

BEGIN
  --
  --*******************************************************
  -- process that generates the WAVES waveform
  --*******************************************************

       WAVES: waveform(wpl);

  --*******************************************************
  -- processes that convert the WPL values to 1164 Logic Values
  --*******************************************************

  WAV_STIM_CLK                <= STIM_1164(wpl.wpl( 1 ));
  WAV_STIM_DACK               <= STIM_1164(wpl.wpl( 2 ));
  WAV_STIM_IN_PORT            <= STIM_1164(wpl.wpl( 3 to 10 ));
  WAV_EXPECT_R_W              <= EXPECT_1164(wpl.wpl( 11 ));
  WAV_EXPECT_AS               <= EXPECT_1164(wpl.wpl( 12 ));
  WAV_EXPECT_ADDR_BUS         <= EXPECT_1164(wpl.wpl(13 to 20 ));
  WAV_EXPECT_READ_FINISH      <= EXPECT_1164(wpl.wpl( 21 ));

  --********************************************
  -- UUT Port Map - Name Semantics Denote Usage
  --********************************************

u1: micro_p
  port map(
       CLK             => WAV_STIM_CLK,
       DACK            => WAV_STIM_DACK,
```

```
    In_port         => WAV_STIM_IN_PORT,
    R_W             => ACTUAL_R_W,
    AS              => ACTUAL_AS,
    ADDR_BUS        => ACTUAL_ADDR_BUS,
    READ_FINISH     => ACTUAL_READ_FINISH);


--***********************************************************
-- Monitor Processes To Verify The UUT Operational Response
--***********************************************************

Monitor_ADDR_BUS:
  PROCESS(ACTUAL_ADDR_BUS, WAV_expect_ADDR_BUS)
  BEGIN
       assert(Compatible (actual => ACTUAL_ADDR_BUS,
                          expected => WAV_expect_ADDR_BUS))
       report "Error on address bus" severity WARNING;

  IF ( Compatible ( ACTUAL_ADDR_BUS,
                    WAV_expect_ADDR_BUS) ) THEN
    FAIL_SIGNAL <='L'; ELSE FAIL_SIGNAL <='1';
  END IF;
  END PROCESS;


Monitor_R_W:
  PROCESS(ACTUAL_R_W, WAV_expect_R_W)
  BEGIN
       assert(Compatible (actual => ACTUAL_R_W,
                          expected => WAV_expect_R_W))
       report "Error on R_W output" severity WARNING;

  IF ( Compatible ( ACTUAL_R_W,    WAV_expect_R_W) ) THEN
    FAIL_SIGNAL <='L'; ELSE FAIL_SIGNAL <='1';
  END IF;
  END PROCESS;

Monitor_AS:
  PROCESS(ACTUAL_AS, WAV_expect_AS)
  BEGIN
       assert(Compatible (actual => ACTUAL_AS,
                          expected => WAV_expect_AS))
       report "Error on AS output" severity WARNING;

  IF ( Compatible ( ACTUAL_AS,    WAV_expect_AS) ) THEN
    FAIL_SIGNAL <='L'; ELSE FAIL_SIGNAL <='1';
  END IF;
```

```
   END PROCESS;

 Monitor_READ_FINISH:
   PROCESS(ACTUAL_READ_FINISH, WAV_expect_READ_FINISH)
   BEGIN
        assert(Compatible (actual => ACTUAL_READ_FINISH,
                           expected => WAV_expect_READ_FINISH))
        report "Error on READ_FINISH output" severity WARNING;

   IF ( Compatible ( ACTUAL_READ_FINISH,
                     WAV_expect_READ_FINISH) ) THEN
     FAIL_SIGNAL <='L'; ELSE FAIL_SIGNAL <='1';
   END IF;
   END PROCESS;

END micro_p_test;
```

Figure 8-13. Testbench for the Embedded Microprocessor Example

In ***summary of testbench generation***, we have described the structure of a typical testbench which consists of seven essential elements: context clauses, UUT component and configuration declaration, connection signal declaration, invocation of waveform generator, translation functions, UUT component instantiation, and monitor processes. This testbench was created to support the relative edge placement in the WAVES.

In ***summary of the Relative Edge Placement Example***, we have presented a systematic approach in creating complete WAVES and VHDL files required to carry out the simulation. This systematic approach consists of three principal steps: developing VHDL models for the UUT, generating the WAVES dataset, and generating the testbench. We also demonstrated WAVES' capability in handling the relative timing and edge transitions which are frequently used in normal device specifications. In the next section, we present another example that involves another complex timing issue: the bi-directional nature of a device.

8.2 Bi-Directional Pin Issues in WAVES

A reasonable system design project will utilize one or more devices with bi-directional pins, such as microprocessors, transceivers, or shift registers. This implies that WAVES must be able to specify test vectors for these bi-directional devices. All the examples we presented so far did not have bi-directional pins and required only a single time set or frame set array to describe a complete waveform. In general, however, a device with bi-directional pins requires multiple time sets in describing the bi-directional nature of the pins. Here, we will consider a device with multiple bi-directional pins which introduces a new challenge to WAVES. We will use an 8-Input Universal Shift/Storage Register (54LS/74LS299) to illustrate WAVES' capability in handling the bi-directional pins.

Using this example, we demonstrate how the WAVES dataset and testbench generation process are affected when the model contains bi-directional pins. We again follow the three distinct logical steps described in Section 6.3 of Chapter 6: developing VHDL models for the UUT, generating the WAVES dataset, and generating the testbench. For this example, we create two sets of pin codes in the waveform generator procedure to handle the bi-directional pins. This implies that the external file must specify a time set to be used for each slice. If you recall from Chapter 7, the external file can contain an integer as a timing command. This integer indicates a timing set selection for each slice. We will further illustrate this aspect of the external file in Section 8.2.3.5.

Here, we begin with the device specifications which describe the functionality and input/output interfaces. Then, we develop all necessary WAVES and VHDL files to support simulation of the WAVES-VHDL design. All source code files for this example are included on the companion CD-ROM in a directory called "***Register***".

8.2.1 Device Specifications for the 54LS/74LS299 Device

This register was modeled after the 54LS/74LS299 series taken from the Fairchild TTL Data Book printed in December of 1978. The 54LS/74LS299 8-input universal shift/storage register, with common parallel input/output pins, is an 8-bit universal shift/storage register with 3-state outputs. Four modes of operation are possible: hold(store), shift left, shift right, and load data. The parallel load inputs and flip-flop outputs are multiplexed to reduce the total number of package pins. Separate outputs are provided for flip-flops, Q0 and Q7, to allow easy cascading. A separate active LOW Master Reset is used to reset the register.

Figure 8-14 lists the register's pin numbers, the names used in the model, pin names, and a short description of each pin's functionality. The truth table for this device is shown in Figure 8-15. For more information including the electrical characteristics, we may refer to the Fairchild TTL Data Book.

Pin Number	Name in Model	Pin Name	Description
1	selection(0)	S0	Mode Selection Input 0
2	enable_out(0)	NOT OE1	3-State Output Enable Input 1
3	enable_out(1)	NOT OE2	3-State Output Enable Input 2
4	IO(6)	IO6	Parallel Data Input/ 3-State Parallel Output 6
5	IO(4)	IO4	Parallel Data Input/ 3-State Parallel Output 4
6	IO(2)	IO2	Parallel Data Input/ 3-State Parallel Output 2
7	IO(0)	IO0	Parallel Data Input/ 3-State Parallel Output 0
8	OUT_0	Q0	Serial Output 0
9	Master_Reset	NOT MR	Asynchronous Master Reset Input (Active LOW)
10	---------	GND	Ground
11	Data_0	DS0	Serial Data Input for Right Shift
12	Clock	CP	Clock
13	IO(1)	IO1	Parallel Data Input/ 3-State Parallel Output 1
14	IO(3)	IO3	Parallel Data Input/ 3-State Parallel Output 3
15	IO(5)	IO5	Parallel Data Input/ 3-State Parallel Output 5
16	IO(7)	IO7	Parallel Data Input/ 3-State Parallel Output 7
17	OUT_7	Q7	Serial Output 7
18	Data_7	DS7	Serial Data Input for Left Shift
19	selection(1)	S1	Mode Selection Input 1
20	----------	VCC	SUPPLY VOLTAGE

Figure 8-14. 8-Input Universal Shift/Storage Register Pin Labels.

The type of operation to be performed is determined by the S0 and S1 selection pins as we illustrate in Figure 8-15. All flip-flop outputs are brought out through tri-state buffers to separate I/O pins that also serve as data inputs in the parallel load mode. Q0 and Q7 are also brought out on other pins for expansion in serial shifting of longer words. A LOW signal on the master reset pin overrides the select and clock pulse inputs and resets the flip-flops. All other state changes are initiated by the rising edge of the clock. Inputs can change when the clock is in either state, provided that the recommended setup and hold times relative to the rising edge of the clock pulse are not violated. A high signal on either of the output enable pins disables the tri-state buffers, and puts the I/O pins in the high impedance state. In this condition, the shift, hold, load and reset operations can still occur. The tri-state buffers are also disabled by high signals on both S0 and S1 in preparation for a parallel load operation.

INPUTS				RESPONSE
MR	S1	S0	CP	
L	X	X	X	Asynchronous Reset: Q0 - Q7= LOW
H	H	H	↑	Parallel Load: I/On → Qn
H	L	H	↑	Shift Right: DS0 → Q0, Q0 → Q1, etc.
H	H	L	↑	Shift Left: DS0 → Q7, Q7 → Q6, etc.
H	L	L	X	Hold

Figure 8-15. 8-Input Universal Shift/Storage Register Truth Table.

In ***summary of the device specifications for 54LS/74LS299***, we have presented a brief description of the functionality and input/output interfaces for the 8-Input Universal Shift/Storage Register through the device specifications. These device specifications are used next to create a VHDL model and the WAVES dataset to support the WAVES-VHDL design simulation.

8.2.2 VHDL Model (UUT) for the 54LS/74LS299 Device

Again, the first step toward the simulation of a WAVES-VHDL design is developing a VHDL model for the UUT. We developed the behavioral model of the 8-Input Universal Shift/Storage Register based on the device specifications, and it is given in Figure 8-16. Adequate comments are added to capture additional useful information. In this example, we introduce more advanced use of VHDL constructs in describing device functionality. For instance, many of the pre-defined attributes such as '**Last_Event**, 'Stable, and 'Event are used to implement extensive time-checking routines. The timing checks include pulse width checks, setup checks, and hold checks. In addition, the behavior of the shift register is implemented using the guarded block statement. For more information regarding the pre-defined attributes and the guarded block statement, we may consult the IEEE-1076 standard. The actual VHDL code for the shift register is included on the companion CD-ROM with the file name "***Register.Vhd***" in the *Register* directory.

```
-- TITLE: FAIRCHILD 8-input universal shift/storage register;
--            54LS/74LS299
-- DATE : 16 June 1995
--
-- VERSION : 2.0
-- FILENAME : register_with_timing.vhd
-- FUNCTION : Entity and architecture for 54LS/74LS299;
--            '299 is 8-bit universal
-- shift/storage
-- register with 3-state outputs. Four modes of operation are
-- possible: hold(store),shift left, shift right and load data.
-- The parallel load inputs and flip-flop outputs are
-- multiplexed to reduce total number of package pins.
-- Separate outputs are provided for flip-flops Q0 and Q7
-- to allow easy cascading. A separate active
-- low master reset is used to reset the register.
--
-- Dataflow Model.
--
-- AUTHOR       : John Murphy
-- ORGANIZATION: Company XYZ
--
-- HISTORY:
-- 28 Apr 95 - v1.0 -  Initial version, functional only,
-- no timing.
-- 31 May 95 - v2.0 -  Final version with worst case timing.
--

library ieee;
use ieee.std_logic_1164.all;

entity shift_register IS
-- Timings are passed through the generic
GENERIC( to_min : TIME := 2 NS; clk_to_IO : TIME := 32 NS;
             clk_to_Q0_or_7 : TIME := 34 NS;
             Reset_to_OUTPUTs : TIME := 36 NS;
             out_enable_Z_to_HI : TIME := 25 NS;
             out_enable_HI_to_Z : TIME := 27 NS;
             out_enable_Z_to_LO : TIME := 30 NS;
             out_enable_LO_to_Z : TIME := 34 NS;
             Setup_Data_to_clk : TIME := 30 NS;
             Setup_selection_to_clk : TIME := 41 NS;
```

```
            Hold_all_inputs_to_clk : TIME := 0 NS;
            Reset_to_clock_hi : TIME := 5 NS;
            Reset_lo_pulse_width : TIME := 22 NS;
            clk_hi_pulse_width : TIME := 30 NS);

PORT( selection, enable_out : IN STD_LOGIC_VECTOR(0 TO 1);
      Clock,Data_0,Data_7,Master_Reset : IN STD_LOGIC;
      IO : INOUT STD_LOGIC_VECTOR(0 TO 7) BUS;
      OUT_0,OUT_7 : OUT STD_LOGIC);

BEGIN

-- Master reset width check
    ASSERT
      (NOT ((Master_Reset = '0' AND NOT Master_Reset'STABLE) AND
        Master_Reset'last_event >= Reset_lo_pulse_width))
       REPORT
       "RESET PULSE WIDTH TOO SHORT"
     SEVERITY WARNING;
-- Minimum clock width pulse check
     ASSERT
       (NOT ((Clock='1' AND NOT Clock'STABLE) AND
        Clock'last_event >= clk_hi_pulse_width))
       REPORT
       "CLOCK PULSE WIDTH TOO SHORT"
     SEVERITY WARNING;

-- Setup timing checks for selection, IO, Data_0 and Data_7
     ASSERT
       (NOT (Clock = '1' AND Clock'EVENT AND NOT
              selection'STABLE(Setup_selection_to_clk)))
     REPORT
       "Setup TIME VIOLATION ON selection PINS"
     SEVERITY WARNING;

     ASSERT
       (NOT (Clock = '1' AND Clock'EVENT AND NOT
             IO'STABLE(Setup_Data_to_clk) AND
             selection ="11"))
     REPORT
       "Setup TIME VIOLATION ON IO PINS"
     SEVERITY WARNING;
     ASSERT
       (NOT (Clock = '1' AND Clock'EVENT AND NOT
             Data_0'STABLE(Setup_Data_to_clk)))
     REPORT
       "Setup TIME VIOLATION ON Data_0 PIN"
```

```
      SEVERITY WARNING;

      ASSERT
        (NOT (Clock = '1' AND Clock'EVENT AND NOT
              Data_7'STABLE(Setup_Data_to_clk)))
      REPORT
        "Setup TIME VIOLATION ON Data_7 PIN"
      SEVERITY WARNING;

-- Hold timing checks for IO,
       ASSERT
        (NOT (Clock = '1' AND IO'EVENT AND NOT
              Clock'STABLE(Hold_all_inputs_to_clk)))
      REPORT
        "HOLD TIME VIOLATION ON IO PINS"
      SEVERITY WARNING;

      ASSERT
        (NOT (Clock = '1' AND selection'EVENT AND NOT
              Clock'STABLE(Hold_all_inputs_to_clk)))
      REPORT
        "HOLD TIME VIOLATION ON selection PINS"
      SEVERITY WARNING;

      ASSERT
        (NOT (Clock = '1' AND Data_0'EVENT AND NOT
              Clock'STABLE(Hold_all_inputs_to_clk)))
      REPORT
        "HOLD TIME VIOLATION ON Data_0 PIN"
      SEVERITY WARNING;

      ASSERT
        (NOT (Clock = '1' AND Data_7'EVENT AND NOT
              Clock'STABLE(Hold_all_inputs_to_clk)))
      REPORT
        "HOLD TIME VIOLATION ON Data_7 PIN"
      SEVERITY WARNING;

END shift_register;

architecture behavioral of  shift_register IS
  SIGNAL Q :  STD_LOGIC_VECTOR(0 TO 7) REGISTER;

BEGIN

 Memory_Reset: BLOCK ( Master_Reset = '0')
 begin
```

```
     Q <=  GUARDED "00000000";
 END BLOCK;

 CLOCKING: BLOCK(Clock = '1' AND NOT Clock'STABLE AND
Master_Reset= '1')
   BEGIN
     Shift_right: BLOCK(selection = "10" AND GUARD)
       BEGIN
         Q <= GUARDED Data_0 & Q(0 TO 6);
       END BLOCK Shift_right;
--
     Shift_left: BLOCK(selection = "01" AND GUARD)
       BEGIN
         Q <= GUARDED Q(1 TO 7) & Data_7;
       END BLOCK Shift_left;
--
    Parrel_load : BLOCK(selection = "11" AND GUARD)
      BEGIN
        Q <= GUARDED IO;
      END BLOCK Parrel_load;

  END BLOCK CLOCKING;

  CHECK_OE : BLOCK ( enable_out = "00"  )
   BEGIN
    IO <= GUARDED Q AFTER Reset_to_OUTPUTs WHEN Q = "00000000"
ELSE
                   Q AFTER out_enable_Z_to_HI WHEN IO =
"ZZZZZZZZ" ELSE
                   Q AFTER clk_to_IO;
   END BLOCK CHECK_OE;

  CHECK_Z : BLOCK(TRUE)
   BEGIN
     IO(0) <= GUARDED 'Z' AFTER out_enable_HI_to_Z
                     WHEN IO(0) = '1' ELSE
                     'Z' AFTER out_enable_LO_to_Z;
     IO(1) <= GUARDED 'Z' AFTER out_enable_HI_to_Z
                     WHEN IO(0) = '1' ELSE
                     'Z' AFTER out_enable_LO_to_Z;
     IO(2) <= GUARDED 'Z' AFTER out_enable_HI_to_Z
                     WHEN IO(0) = '1' ELSE
                     'Z' AFTER out_enable_LO_to_Z;
     IO(3) <= GUARDED 'Z' AFTER out_enable_HI_to_Z
                     WHEN IO(0) = '1' ELSE
                     'Z' AFTER out_enable_LO_to_Z;
     IO(4) <= GUARDED 'Z' AFTER out_enable_HI_to_Z
```

```
                    WHEN IO(0) = '1' ELSE
                    'Z' AFTER out_enable_LO_to_Z;
    IO(5) <= GUARDED 'Z' AFTER out_enable_HI_to_Z
                    WHEN IO(0) = '1' ELSE
                    'Z' AFTER out_enable_LO_to_Z;
    IO(6) <= GUARDED 'Z' AFTER out_enable_HI_to_Z
                    WHEN IO(0) = '1' ELSE
                    'Z' AFTER out_enable_LO_to_Z;
    IO(7) <= GUARDED 'Z' AFTER out_enable_HI_to_Z
                    WHEN IO(0) = '1' ELSE
                    'Z' AFTER out_enable_LO_to_Z;
  END BLOCK CHECK_Z;

  OUT_0 <= Q(0) AFTER clk_to_Q0_or_7;
  OUT_7 <= Q(7) AFTER clk_to_Q0_or_7;

END Behavioral;
```

Figure 8-16. Behavioral Model for the Device 54LS/74LS299

In ***summary of the VHDL model generation***, we have demonstrated some advanced VHDL constructs in creating an efficient and readable VHDL device model based on the device specifications. Even though teaching VHDL modeling techniques is not in the scope of this text, we portray some useful techniques in this example, for modeling the bi-directional device. Next, we generate the WAVES dataset to support the WAVES-VHDL design simulation.

8.2.3 WAVES Dataset Generation for the 54LS/74LS299 Device

Our second principal step, required to carry out the WAVES-VHDL simulation, is to generate a corresponding WAVES dataset for the UUT. The WAVES dataset development procedure consists of five sub-tasks: creating of a header file for documentation, creating and compiling the test pins package, modifying and compiling the WAVES Objects package, creating and compiling the waveform generator package, and creating an external (test vector) file. Here, we present each WAVES dataset element generated by each sub-task. We begin with the header file.

8.2.3.1 Header File for the 54LS/74LS299 Device

The header file captures information that is necessary and useful to describe the WAVES dataset completely. Such information includes the dataset identification and construction information, external test vector file identification, and waveform generator procedure identification. The header file for the 8-bit universal shift/storage register is shown in Figure 8-17.

```
-- ************************************************
--
-- ******** Header File for Entity: shift_register
--
-- ************************************************
--
-- Data Set Identification Information
--
TITLE          A General Description
DEVICE_ID      shift_register

DATE           Fri Jul 28 10:44:44 1995
ORIGIN         Company X Design Team
AUTHOR         Company or Person
AUTHOR         Maybe Multiple ... Companies or People
DATE           Fri Jul 28 10:44:44 1995
ORIGIN         Modified by Company X Design Team
AUTHOR         Who did it Company or Person

OTHER          Use TestBench register_tstbench.vhd
OTHER          Any general comments you want
OTHER          Built Using the WAVES-VHDL 1164 STD Libraries
--
-- Data Set Construction Information
--
WAVES_FILENAME   register_pins.vhd            WORK
library          WAVES_1164;
use              WAVES_1164.WAVES_1164_Pin_Codes.all;
use              WAVES_1164.WAVES_1164_Logic_Value.all;
use              WAVES_1164.WAVES_Interface.all;
use              WORK.UUT_Test_pins.all;
WAVES_UNIT       WAVES_OBJECTS                WORK
WAVES_FILENAME   register_wgen.vhd            WORK
--
-- external test vector file identification
EXTERNAL_FILENAME   vectors.txt            VECTORS
--
-- waveform generator procedure identification
WAVEFORM_GENERATOR_PROCEDURE
WORK.waves_shift_register.waveform
```

Figure 8-17. The Header File for 54LS/74LS299

8.2.3.2 Test Pins Package for the 54LS/74LS299 Device

After the header file is defined, we need to define the **Test_Pins** type. The waveform information of the UUT is applied and observed through the WAVES port list and the **Test_Pins** type is used to define the WAVES port list. The **Test_Pins** declaration for the 8-bit universal shift/storage register is given in Figure 8-18.

```
--
PACKAGE uut_test_pins IS
TYPE test_pins IS (selection_0, selection_1, enable_out_0,
    enable_out_1, Clock, Data_0, Data_7, Master_Reset,
    IO_0, IO_1, IO_2, IO_3, IO_4, IO_5, IO_6, IO_7,
    OUT_0, OUT_7);
END uut_test_pins;
```

Figure 8-18. Test Pins Package for the 54LS/74LS299 Device

8.2.3.3 Waves_Objects Package for the 54LS/74LS299 Device

Next, we need to add proper context clauses to the beginning of the **Waves_Objects** package and compile this package into the work library. The context clauses provide visibility to the **Waves_Interface**, **Logic_Value**, **Pin_Codes**, and **Test_Pins** packages. The context clauses for the 8-bit universal shift/storage register are presented in Figure 8-19.

```
use STD.TEXTIO.all;
library WAVES_STD;
use WAVES_STD.WAVES_SYSTEM;

-- A context clause providing visibility to an analyzed copy of
-- WAVES_INTERFACE is required at this point.
-- Context clauses providing visibility to LOGIC VALUE,
-- TEST PINS, and PIN CODES are required at this point.
Library WAVES_1164;
use WAVES_1164.WAVES_Interface.all;
use WAVES_1164.WAVES_1164_Logic_Value.all;
use WAVES_1164.WAVES_1164_Pin_Codes.all;
use WORK.UUT_Test_pins.all;
```

Figure 8-19. The Context Clauses for the Waves_Objects Package

8.2.3.4 Waveform Generator Procedure for the 54LS/74LS299 Device

Next, we define the waveform generator procedure. The waveform generator procedure reads a slice of pin codes from an external test vector file. Then, it constructs a slice of the waveform utilizing the definition of the **Frame_Set_Array** declaration, the outputs stimulus, and the expected responses to the testbench through the WAVES port list.

As we mentioned previously, this device has multiple bi-directional pins which introduces complex timing issues. To handle this bi-directional nature of the pins, we define multiple pin code sets. We create two sets of pin codes in the WAVES waveform generator procedure which implies that the external file needs to specify which set is to be used for each slice. Now, let's look at the WGP for the 8-bit universal shift/storage register, shown in Figure 8-20.

```
-- Context Clauses

LIBRARY WAVES_STD;
USE WAVES_STD.WAVES_Standard.all;
library WAVES_1164;
USE STD.textio.all;
USE WAVES_1164.waves_1164_frames.all;
USE WAVES_1164.waves_1164_pin_codes.all;
USE WAVES_1164.waves_interface.all;
USE work.waves_objects.all;
USE work.uut_test_Pins.all;

PACKAGE WGP_shift_register is
     PROCEDURE  waveform(SIGNAL WPL : inout WAVES_PORT_LIST);
END WGP_shift_register;

-------------------------------------------------------

PACKAGE BODY WGP_shift_register is

-- This is the uut pin declaration pin and ordering
-- Remember you need to match the External file to This order
--
--selection_0, selection_1, enable_out_0, enable_out_1, Clock,
-- Data_0, Data_7, Master_Reset, IO_0, IO_1, IO_2, IO_3,
-- IO_4, IO_5, IO_6, IO_7, OUT_0, OUT_7

    PROCEDURE  waveform(SIGNAL WPL : inout WAVES_PORT_LIST) is

       FILE vector_file : TEXT is in "VECTORS.TXT";
```

```
    VARIABLE vector : FILE_SLICE := NEW_FILE_SLICE;

--  Grouping of the pins

    CONSTANT selection: pinset:= new_pinset(( selection_0,
                                              selection_1));

    CONSTANT enable_out: pinset:= new_pinset(( enable_out_0,
                                              enable_out_1));

    CONSTANT IO: pinset:= new_pinset(( IO_0, IO_1, IO_2,
                        IO_3, IO_4, IO_5, IO_6, IO_7));

     CONSTANT outputs: pinset:= new_pinset((OUT_0, OUT_7));

     CONSTANT in_pins: pinset:= new_pinset((Data_0, Data_7,
                                         Master_Reset));

     CONSTANT inputs: pinset:= in_pins or selection or
                                               enable_out;

   -- Declare The Frame Sets (multiple timing sets)
   --
    Variable WTL : Wave_timing_list (1 to 2) := (
    --
    -- Frame Set 1 for shift and hold operations
    -- The bi-directional pins acts as output pins
    -- in these modes
    --
     (  Delay  => Delay ( 100 ns ),
     Timing => New_Time_Data(
     New_frame_set_array(Pulse_high( 50 ns, 80 ns), Clock) +
         New_frame_set_array(Window( 85 ns, 95 ns), IO) +
         New_frame_set_array(Non_return( 5 ns), inputs) +
         New_frame_set_array(window( 85 ns, 95 ns), outputs)
     )),
    --
    -- Frame Set 2 for load operation
    -- The bi-directional pins acts as input pins in
    -- this mode
     --
     (  Delay  => Delay ( 100 ns ),
     Timing => New_Time_Data(
     New_frame_set_array(Pulse_high( 50 ns, 80 ns), Clock)+
     New_frame_set_array(Non_return( 5 ns), IO) +
     New_frame_set_array(Non_return( 5 ns), inputs) +
```

```
        New_frame_set_array(window( 85 ns, 95 ns), outputs)
        )));

    BEGIN
      loop
        -- get first vector
        READ_FILE_SLICE (vector_file, Vector);
        exit when vector.end_of_file;
       apply(wpl, vector.codes.all, WTL(vector.fs_integer));
      end loop;

    END waveform;

END WGP_shift_register;
```

Figure 8-20. Waveform Generator Procedure for the 54LS/74LS299 Device

As usual, the WGP requires visibility of the **Waves_Standard**, **Textio**, **Waves_1164_Frames**, **Waves_1164_Pin_Codes**, **Waves_Interface**, **Waves_Objects**, and **UUT_Test_Pins** packages. The visibility to these packages is ensured by adding appropriate context clauses to the beginning of the WGP package.

The name of the external test vector file that contains the pin codes is "vectors.txt" and it is available in the register directory on the companion CD-ROM. The content of the external file will be presented in Section 8.2.3.5. Next, we group input and output signals based on their timing characteristics and the direction of each signal (defining the pinsets). This pin grouping simplifies the frame set array declaration because the grouping allows us to use the same frame set for the pins with the same timing and electrical characteristics.

Next, we declare the necessary time sets that are sufficient to describe the waveform for the UUT. Unlike all the examples we presented previously, this example requires two separate time sets to handle the complex timing issues caused by the bi-directional nature of the pins. As a result, the **Waves_Timing_List** type, which is defined in the **Waves_Standard** package, is used to define multiple time sets. There are eight bi-directional pins (**IO_0** through **IO_7**) on this device. As described in the device specifications, these pins act as output pins during reset, shift (both left and right), and hold operations. However, these pins function as input pins for the load operation. In this example, the frame set 1 (or time set 1) is defined to handle the shift and hold operations, and the frame set 2 (or time set 2) is defined to handle the load operation.

Finally, the WGP reads the external file, one slice of pin codes at a time, into the file slice and constructs each slice of the waveform utilizing the proper time set indicated in the file slice. The construction of the waveform is again accomplished by the APPLY function. Here, we should note that the APPLY function in this example is slightly different than ones we used in previous examples. In particular, it takes only three parameters, one of which is a timing set selection integer, in contrast to the four we used in our previous examples. This means that a different APPLY function is selected and executed depending on the number and type of parameters passed to the function. In VHDL, this type of function is called an overloaded function. In general, the overloaded functions perform the same basic function, such as constructing a slice of waveform in this example; however, there exists more than one body for the function to act on different sets of parameters (data). For more information regarding the overloaded function, we may consult the IEEE 1076-93 Standard. Again, the constructed waveform is provided to the testbench via the **Waves_Port_List** during simulation. This process repeats until the end of the external file is reached.

8.2.3.5 External Test Vector File for the 54LS/74LS299 Device

Test vectors in the external file test all five operations of the 54LS/74LS299; reset, shift right, shift left, hold, and load. The external file format, shown in Figure 8-21, lists the conditions on the pins (selection, output enable, clock, data serial 0, data serial 7, master reset, and the I/O pins, from left to right) followed by the identifier for the proper time set. Time set one sets the I/O pins to be compared (output mode) and time set two drives the I/O as input. (Recall the declaration of the time sets in Figure 8-20.) The first vector resets the register, then six shift right operations are then performed, followed by five shift left operations. The resulting output is then held for five cycles. The register is then loaded with data and the output disabled with the loaded information shifted left. The I/O pins are tested for a tri-state condition and the last operation enables the outputs and performs a final shift left operation. For this example, we could compress the test vectors by utilizing the skip command we described in Chapter 7. However, we avoided using the skip command to improve the readability. We could reformat this external file if we wish to do so, utilizing the various techniques described in Chapter 7, and it would, in fact, be a useful exercise.

```
% ss        dd  m               t
% ee  oo c  aa  r   iiiiiiii    e
% ll  ee l  tt  s   oooooooo qq s
% 01  12 k  01  t   01234567 07 t
%
% clear or reset
```

```
   --  00 -  --  0    00000000  00 : 1 ;

% shift right

   10  00 1  11  1    10000000  10 : 1 ;
   10  00 1  11  1    11000000  10 : 1 ;
   10  00 1  11  1    11100000  10 : 1 ;
   10  00 1  11  1    11110000  10 : 1 ;
   10  00 1  11  1    11111000  10 : 1 ;
   10  00 1  11  1    11111100  10 : 1 ;

% shift left

   01  00 1  11  1    11111001  11 : 1 ;
   01  00 1  11  1    11110011  11 : 1 ;
   01  00 1  11  1    11100111  11 : 1 ;
   01  00 1  11  1    11001111  11 : 1 ;
   01  00 1  10  1    10011110  10 : 1 ;

% hold

   00  00 1  10  1    10011110  10 : 1 ;
   00  00 1  10  1    10011110  10 : 1 ;
   00  00 1  10  1    10011110  10 : 1 ;
   00  00 1  10  1    10011110  10 : 1 ;

% load

   11  10 1  10  1    01010101  01 : 2 ;

% enable & shift

   01  01 1  01  1    ZZZZZZZZ  11 : 1 ;
   01  00 1  01  1    01010111  01 : 1 ;
```

Figure 8-21. External Test Vector File for the 54LS/74LS299 Device

In ***summary of WAVES dataset generation for 54LS/74LS299***, we have demonstrated capability and flexibility of the WAVES dataset in handling the complex timing issues. In particular, we have used multiple time sets to describe the bi-directional nature of test pins. We also explained the purpose and usage of the overloaded functions. In addition, we described the external file format required to support multiple time sets. Now, we are ready to generate a testbench that utilizes both the VHDL model of the UUT and the WAVES dataset to carry out the simulation.

8.2.4 Testbench Generation for the 54LS/74LS299 Device

The testbench generation is the *last principal step* for carrying out the simulation of the WAVES-VHDL design. The testbench establishes communications between the WAVES dataset and the UUT VHDL model, monitors the response of the UUT, and compares that response with the expected response.

As usual, this testbench contains seven distinct elements. They are the context clauses, the UUT component and configuration declarations, the connection signal declaration, invocation of the waveform generator, the translation functions, the UUT component instantiation, and the monitor processes. The purpose and role of each element in the testbench was described in detail in Section 6.3 of Chapter 6. Here, we simply present the testbench for 54LS/74LS299, in Figure 8-22, without a great deal of explanation. However, we do provide brief explanations for the codes that are specifically inserted to handle the bi-directional pins.

```
-- Context clauses
LIBRARY ieee;
USE ieee.std_logic_1164.ALL;

LIBRARY waves_1164;
USE waves_1164.WAVES_1164_utilities.all;

USE WORK.UUT_test_pins.all;
USE work.waves_objects.all;

USE work.WGP_shift_register.all;

ENTITY test_bench IS
END test_bench;


ARCHITECTURE shift_register_test OF test_bench IS


  --****************************************************
  --**********Component Declaration   *******************
  --****************************************************

  COMPONENT shift_register
    PORT ( selection        : IN   std_logic_vector( 0 to  1 );
           enable_out       : IN   std_logic_vector( 0 to  1 );
           Clock            : IN   std_logic;
           Data_0           : IN   std_logic;
```

```
          Data_7              : IN   std_logic;
          Master_Reset        : IN   std_logic;
          IO                  : INOUT std_logic_vector( 0 to  7 );
          OUT_0               : OUT  std_logic;
          OUT_7               : OUT  std_logic);
    END COMPONENT;

-- Configuration statement
FOR ALL:shift_register USE ENTITY
work.shift_register(behavioral);

-- Connections signals


  --*********************************************************
  -- stimulus signals for the waveforms mapped into UUT INPUTS
  --*********************************************************

    SIGNAL WAV_STIM_selection     :std_logic_vector(  0 to  1 );
    SIGNAL WAV_STIM_enable_out    :std_logic_vector(  0 to  1 );
    SIGNAL WAV_STIM_Clock               :std_logic;
    SIGNAL WAV_STIM_Data_0              :std_logic;
    SIGNAL WAV_STIM_Data_7              :std_logic;
    SIGNAL WAV_STIM_Master_Reset        :std_logic;

  --***************************************************
  -- Expected signals used in monitoring the UUT OUTPUTS
  --***************************************************

    SIGNAL FAIL_SIGNAL            :std_logic;
    SIGNAL WAV_EXPECT_IO          :std_ulogic_vector( 0 to  7 );
    SIGNAL WAV_EXPECT_OUT_0             :std_logic;
    SIGNAL WAV_EXPECT_OUT_7             :std_logic;

  --*****************************************************
  -- UUT Output signals used In Monitoring ACTUAL Values
  --*****************************************************

    SIGNAL ACTUAL_OUT_0               :std_logic;
    SIGNAL ACTUAL_OUT_7               :std_logic;

  --********************************************************
  -- Bi_directional signals used  for stimulus signals mapped
  -- into UUT INPUTS and also monitoring the UUT OUTPUTS
  --********************************************************

    SIGNAL BI_DIREC_IO            :std_logic_vector(  0 to  7 );
```

```
--**********************************************************
-- WAVES signals OUTPUTing each slice of the waves port list
--**********************************************************

      SIGNAL wpl  : WAVES_port_list;

BEGIN
  --
  --**********************************************************
  -- process that generates the WAVES waveform
  --*********************************************************

      WAVES: waveform(wpl);

  --********************************************************
  -- processes that translate the WPL values
  --      to 1164 Logic Values
  --*******************************************************

  WAV_STIM_selection          <= STIM_1164(wpl.wpl( 1 to 2 ));
  WAV_STIM_enable_out         <= STIM_1164(wpl.wpl( 3 to 4 ));
  WAV_STIM_Clock              <= STIM_1164(wpl.wpl( 5 ));
  WAV_STIM_Data_0             <= STIM_1164(wpl.wpl( 6 ));
  WAV_STIM_Data_7             <= STIM_1164(wpl.wpl( 7 ));
  WAV_STIM_Master_Reset       <= STIM_1164(wpl.wpl( 8 ));
  BI_DIREC_IO               <= BI_DIR_1164(wpl.wpl( 9 to 16 ));
  WAV_EXPECT_IO             <= EXPECT_1164(wpl.wpl( 9 to 16 ));
  WAV_EXPECT_OUT_0              <= EXPECT_1164(wpl.wpl( 17 ));
  WAV_EXPECT_OUT_7              <= EXPECT_1164(wpl.wpl( 18 ));

  --*******************************************
  -- UUT component instantiation - UUT Port Map
  --*******************************************

  u1: shift_register
  PORT MAP(
    selection           => WAV_STIM_selection,
    enable_out          => WAV_STIM_enable_out,
    Clock               => WAV_STIM_Clock,
    Data_0              => WAV_STIM_Data_0,
    Data_7              => WAV_STIM_Data_7,
    Master_Reset        => WAV_STIM_Master_Reset,
    IO                  => BI_DIREC_IO,
    OUT_0               => ACTUAL_OUT_0,
    OUT_7               => ACTUAL_OUT_7);
```

```
--*********************************************************
-- Monitor Processes To Verify The UUT Operational Response
--*********************************************************

 -- Monitor the Bi-directional IO
Monitor_IO:
  PROCESS(BI_DIREC_IO, WAV_expect_IO)
  BEGIN
       assert(Compatible (actual => BI_DIREC_IO,
                          expected => WAV_expect_IO))
       report "Error on IO output" severity WARNING;

  IF ( Compatible ( BI_DIREC_IO,    WAV_expect_IO)) THEN
    FAIL_SIGNAL <='L'; ELSE FAIL_SIGNAL <='1';
  END IF;
  END PROCESS;

Monitor_OUT_0:
  PROCESS(ACTUAL_OUT_0, WAV_expect_OUT_0)
  BEGIN
       assert(Compatible (actual => ACTUAL_OUT_0,
                          expected => WAV_expect_OUT_0))
       report "Error on OUT_0 output" severity WARNING;

  IF ( Compatible ( ACTUAL_OUT_0,    WAV_expect_OUT_0) ) THEN
    FAIL_SIGNAL <='L'; ELSE FAIL_SIGNAL <='1';
  END IF;
  END PROCESS;

Monitor_OUT_7:
  PROCESS(ACTUAL_OUT_7, WAV_expect_OUT_7)
  BEGIN
       assert(Compatible (actual => ACTUAL_OUT_7,
                          expected => WAV_expect_OUT_7))
       report "Error on OUT_7 output" severity WARNING;

  IF ( Compatible ( ACTUAL_OUT_7,    WAV_expect_OUT_7) ) THEN
    FAIL_SIGNAL <='L'; ELSE FAIL_SIGNAL <='1';
  END IF;
  END PROCESS;

END shift_register_test;
```

Figure 8-22. Testbench for the 54LS/74LS299 Device

There is one signal assignment statement in this testbench specifically dedicated to handle the bi-directional nature of the pins. This signal assignment is shown below:

```
BI_DIREC_IO     <= BI_DIR_1164(wpl.wpl( 9 to 16 ));
```

In VHDL and WAVES, the right hand side of the expression is evaluated first, prior to the actual signal assignment. This means that a translation function **Bi_Dir_1164** is called with an argument wpl.wpl(9 to 16) before the signal assignment. The argument wpl.wpl(9 to 16) represents the bi-directional pins in position within the WAVES Port List. More precisely, this function converts WAVES port integers into an 1164 **Std_Logic** values and these **Std_Logic** values are assigned to the signal **Bi_Direc_IO**. Then, the signal **Bi_Direc_IO** is used to drive or observe the bi-directional pins depending on the mode of the operation. As you recall, we have described this translation function in great detail in Section 6.2 of Chapter 6. In fact, this function is included in the package **Waves_1164_Utilities** which resides in the library **WAVES_1164**. This function was developed specifically to support the bi-directional nature of the pins.

In ***summary of testbench generation for the 8-bit universal shift/storage register***, we have utilized the translation function **Bi_Dir_1164** in the testbench to support the bi-directional pin devices. We also have demonstrated that only minor changes to the testbench are required to support the bi-directional pin devices.

In ***summary of the 8-bit universal shift/storage register example***, we have demonstrated that WAVES is capable of supporting complex timing issues. In particular, we have illustrated the use of the multiple timing sets in handling the bi-directional devices. In addition, we used the systematic approach in creating the WAVES dataset and testbench to support the WAVES-VHDL simulation.

In ***Summary of Chapter 8,*** we began this chapter with some realistic, practical, and complex timing issues which presented a new challenge to WAVES. In particular, we first considered issues involving the relative timing and edge transitions among signals and provided the WAVES solution to this issue. We also considered issues involving the bi-directional pin devices and presented the WAVES solution. With experience and knowledge gained from these examples, we can now apply WAVES to support realistic, practical design and testing environments. Now, let's revisit the original purpose of WAVES. We recall from Chapter 2 that WAVES is intended to provide a standard way to capture the testing data and facilitate communications between test and design areas. However, all the examples we provided so far were devoted to supporting design simulation environments. In Chapter 9, we provide a case study which demonstrates WAVES' capability to capture existing tester data (such as for ATE support). There, we illustrate that WAVES can be a useful tool in facilitating interfaces between design and testing areas.

CHAPTER 9. CAPTURING WAVEFORMS AND SUPPORTING AUTOMATIC TEST EQUIPMENT (ATE)

Completing the simulation-hardware connection

Given the ability to use WAVES and VHDL, and to handle realistic and practical examples, we may wish to capture some known or existing waveform for use in our WAVES and VHDL simulation. After all, the primary purpose of WAVES is to enhance information exchange between a design environment and a test environment. For example, a hardware-generated waveform is a precise manifestation of some complex timing relationships among logic levels. Given that we are designing some systems which will accept this waveform, we may wish to capture it into a WAVES description, preferably not by hand-crafting the entire, complex sequence. Similarly, we may wish to configure a hardware tester to produce the same waveform we have used in our WAVES dataset, to stimulate the hardware end-product of our design in the same manner we stimulated our VHDL simulation.

As we can see, there are numerous applications that can benefit from smooth information exchanges across the design environment and application domain. In this chapter, we present a case study in which we utilize WAVES as a mechanism for exchanging information between a test and a design environment. In this case study, we create a WAVES dataset from existing tester data which was used to test actual hardware (the AM2901, bit-slice processor). This WAVES dataset may be used to improve the existing design or to emulate the design using a newer technology. We begin with a description of the case study to clarify the use of WAVES as an information exchange mechanism. Then, we create the WAVES dataset from the actual tester data. In the process of creating the WAVES dataset, we will explain some of the files that are specific to the tester, and where to get the necessary information for the dataset.

9.1 Description of the Case Study

The purpose of this case study is to illustrate how WAVES is a useful tool for exchanging information between a design and a test environment, as shown in Figure 9-1. In particular, we will show how WAVES can be used to represent ATE test vectors. In order to initiate our study, the first task was to find or develop an actual ATE test vector, which appeared to be somewhat challenging. Fortunately, our colleagues at the U.S. Air Force Rome Laboratory have developed many test programs for various microelectronics devices and they were kind enough to provide us a set of tester data (files) for this study.

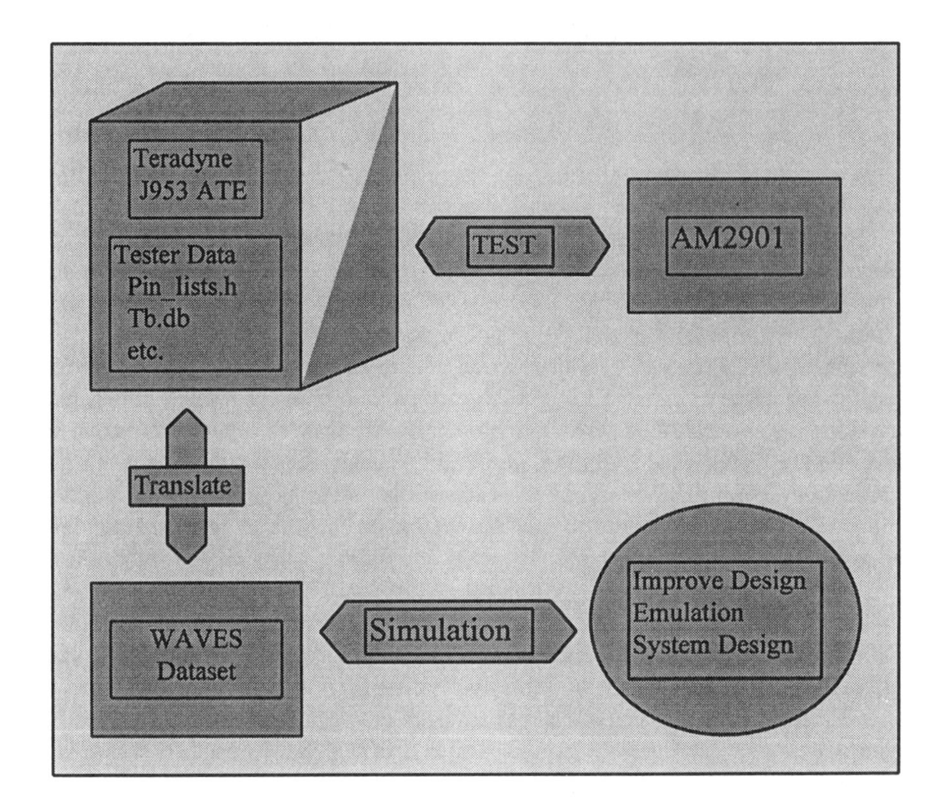

Figure 9-1. WAVES for Exchanging Information

The tester data used in this study was developed by Rome Laboratory to test a 4-bit-wide, bit-slice processor (the AM2901). The ATE, which conducts the actual testing of the AM2901, was the Teradyne J953 system. This means that all tester files, including test vectors used for this study, are specific to the Teradyne J953 ATE. This ATE dependency of the tester data can be easily justified since different ATEs may use different ATE pin codes, formats, vectorization, and vector compression features such as repeat and loop. This brings up an important point that we must realize. In order to create a WAVES dataset which represents ATE test data, we must have a clear understanding of the ATE-specific test data, in addition to our WAVES knowledge. For example, we need to know which tester file to look at if we want to extract information for defining the **Test_Pins** type in the WAVES dataset. From this we can see that the tester data of the Teradyne J953 ATE consists of multiple files. We will introduce these tester files in the next section, when appropriate. Before we begin with this study, we would like to make one more point. Here, we will concentrate upon capturing the ATE tester data using WAVES, which means that we will not spend much time explaining the functionality of the AM2901 and its test vectors.

In ***summary of our description of the case study,*** having discussed the objective, tester specific issues, and approach of our study, now we are ready to construct a WAVES dataset from the Teradyne J953 tester data, for the AM2901. In the next section, as we do so, we will essentially follow the systematic approach we described in Chapter 5.

9.2 A WAVES Dataset for the AM2901

As we described in Chapter 5, the WAVES dataset development procedure which utilizes the **WAVES_1164** library consists of five sub-tasks: creating a header file for documentation, creating and compiling the test pins package, modifying and compiling the WAVES Objects package, creating and compiling the waveform generator package, and creating an external test vector file. Here, we need to generate the WAVES dataset elements using information extracted from the tester data (files) on the Teradyne J953 ATE, which utilizes multiple files to organize the tester data. For example, it uses a file called "**Pin_Lists.H**" to capture test pins and pin grouping information. When we create each element of the WAVES dataset, we will introduce the appropriate tester file at that time. Now, we begin with the test pins package, deferring our discussions of the header file to the end of this section.

9.2.1 The Test Pins Package for the AM2901

The **Test_Pins** package is the first element of WAVES dataset that needs to be defined. As we know, the **Test_Pins** type defines all the input and output pins of the unit under test. The Teradyne J953 ATE keeps information regarding input and output pins of a UUT in a file called "**Pin_Lists.H**". The file **Pin_Lists.H** is shown in Figure 9-2. The first five lines of the file define the input and output interface of the UUT. From this information, we can conclude that AM2901 has 24 input pins, 4 bi-directional pins, and 10 output pins. Using this information, we created the test pins package, and it is shown in Figure 9-3. In addition, the **Pin_Lists.H** file contains pin grouping information such as **Pin_Group_001**, **Pin_Group_011**, and so forth. This pin grouping information will be used in the waveform generator procedure. Note that the **Pin_Lists.H** file is specific to the ATE Teradyne J953. Different ATEs may use different files for the same information.

```
INPUT_PINS = I8 I7 I6 I5 I4 I3 I2 I1 I0 A3 A2 A1
 A0 B3 B2 B1 B0 D3 D2 D1 D0 CN CP RAM3 Q3 RAM0 Q0 OE;
IO_PINS = RAM3 Q3 RAM0 Q0;
OUTPUT_PINS = RAM3 Q3 RAM0 Q0 Y3 Y2 Y1 Y0 P G CN4 OVR Z F3;
OUTPUT_ONLY_PINS = Y3 Y2 Y1 Y0 P G CN4 OVR Z F3;
GND_PINS = GND;
VDD_PINS = VDD;

IDD_TEST_LO_PINS = I8 I6 I4 I2 I0 A2 A0 B2 B0 D2 D0 CP Q3 Q0;
IDD_TEST_HI_PINS = I7 I5 I3 I1 A3 A1 B3 B1 D3 D1 CN RAM3 RAM0
OE;
ALL_PINS = INPUT_PINS OUTPUT_ONLY_PINS;

pin_list_5 =  I8 I7 I6 I5 I4 I3 I2 I1 I0 A3;
pin_list_6 =  A2 A1 A0 B3 B2 B1 B0 D3 D2 D1;
pin_list_7 =  D0 CN CP RAM3 RAM0 Q3 Q0 OE Y3 Y2;
pin_list_8 =  Y1 Y0  P  G CN4 OVR  Z F3;
pin_group_001 =  I8 I7 I6 I5 I4 I3 I2 I1 I0 A3 A2 A1 A0 B3 B2 B1


B0 D3 D2 D1 D0 CN OE;
pin_group_011 =  RAM3 Q3;
pin_group_010 =  RAM0 Q0;
pin_group_002 =  CP;
pin_group_003 =  Y3 Y2 Y1 Y0;
pin_group_012 =   P  G CN4 OVR  Z F3;

DRIVE3 = RAM3 Q3 RAM0 Q0;

INPUT_PINS3 = I8 I7 I6 I5 I4 I3 I2 I1 I0 A3 A2 A1 A0 B3 B2 B1 B0
D3 D2 D1 D0 CN CP OE;

OUTPUT_PINS3 = Y3 Y2 Y1 Y0 P G CN4 OVR F3;

OUTPUT_PIN_PULL_RES = Z;
```

Figure 9-2. Pin_Lists.H File from the Teradyne J953 ATE

```
package uut_test_pins is

  type test_pins is ( I8, I7, I6, I5, I4, I3, I2, I1, I0,
                      A3, A2, A1, A0, B3, B2, B1, B0, D3, D2, D1, D0,
                      CN, CP, RAM3, RAM0, Q3, Q0, OE, Y3, Y2, Y1, Y0,
                      P, G, CN4, OVR, Z, F3 );

end uut_test_pins;
```

Figure 9-3. Test Pins Package for the AM2901

9.2.2 The Waves_Objects Package for the AM2901

The next step in building our WAVES dataset is to add the proper context clauses to the beginning of the **Waves_Objects** package, unless we are using 1164 packages. The context clauses provide visibility of the **Waves_Interface**, **Logic_Value**, **Pin_Codes**, and **Test_Pins** packages. The appropriate context clauses for the bit slice processor are given in Figure 9-4. Here, we should realize that the **Waves_Objects** package is only applicable in supporting a WAVES-VHDL design simulation. Therefore, the ATE does not keep any information regarding the **Waves_Objects** package.

```
use STD.TEXTIO.all;
library WAVES_STD;
use WAVES_STD.WAVES_SYSTEM;

-- A context clause providing visibility to an analyzed copy of
-- WAVES_INTERFACE is required at this point.
-- Context clauses providing visibility to LOGIC VALUE, TEST
PINS,
-- and PIN CODES are required at this point.
Library WAVES_1164;
use WAVES_1164.WAVES_Interface.all;
use WAVES_1164.WAVES_1164_Logic_Value.all;
use WAVES_1164.WAVES_1164_Pin_Codes.all;
use WORK.UUT_Test_pins.all;
```

Figure 9-4. The Context Clauses for the Waves_Objects Package

9.2.3 A Waveform Generator Procedure (WGP) for the AM2901

Next, we need to define the waveform generator procedure. The waveform generator procedure reads a slice of pin codes from an external test vector file. Then it constructs a slice of the waveform utilizing the definitions of the **Frame_Set_Array** declaration, outputs stimulii, and expected responses to the testbench, through the WAVES port list. Let's begin creating the WGP, utilizing the tester files. We first present the WGP for the AM2901 in Figure 9-5, then provide an explanation regarding where the information used in the WGP came from.

```
LIBRARY WAVES_STD;
USE WAVES_STD.WAVES_Standard.all;

USE STD.textio.all;
LIBRARY WAVES_1164;
USE WAVES_1164.waves_1164_frames.all;
USE WAVES_1164.waves_1164_pin_codes.all;
USE WAVES_1164.waves_interface.all;
USE work.waves_objects.all;
USE work.uut_test_Pins.all;

PACKAGE WGP_am2901 is
     PROCEDURE  waveform(SIGNAL WPL : inout WAVES_PORT_LIST);
END WGP_am2901;

-------------------------------------------------------

PACKAGE BODY WGP_am2901 is
-- This is the uut pin declaration
-- Remember you need to match the External file to This order
--
--I8, I7, I6, I5, I4, I3, I2, I1, I0, A3, A2, A1, A0, B3,
--B2, B1, B0, D3, D2, D1, D0, CN, CP, RAM3, RAM0, Q3, Q0,
--OE, Y3, Y2, Y1, Y0, P, G, CN4, OVR, Z, F3

    PROCEDURE  waveform(SIGNAL WPL : inout WAVES_PORT_LIST) is

       FILE vector_file : TEXT is in "vectors.txt";

       VARIABLE vector : FILE_SLICE := NEW_FILE_SLICE;

         --  Grouping of the pins

       CONSTANT P001 : pinset := new_pinset( (I8, I7, I6, I5,
                         I4, I3, I2, I1, I0, A3, A2, A1, A0, B3,
                         B2, B1, B0, D3, D2, D1, D0 ) );
```

```
CONSTANT P011 : pinset := new_pinset((RAM3, Q3 ) );

CONSTANT P010 : pinset := new_pinset((RAM0, Q0 ) );

CONSTANT P002 : pinset := new_pinset((CP ) );

CONSTANT P003 : pinset := new_pinset((Y3, Y2, Y1, Y0 ) );

CONSTANT P012 : pinset := new_pinset((P, G, CN4, OVR, Z,
                                       F3 ));
  -- Declare The Frame Sets (timing sets)
  --
  Variable WTL : Wave_timing_list (1 to 4) := (
  --
  -- Frame Set 1
  --
  (  Delay  => Delay ( 60 ns ),
     Timing => New_Time_Data(
         New_Frame_Set_Array( Non_Return( 0 ns ), P001) +
         New_Frame_Set_Array( Non_Return( 0 ns ),
                              P011) +
         New_Frame_Set_Array( Non_Return( 0 ns ),
                              P010) +
         New_Frame_Set_Array( Pulse_low( 15 ns, 40 ns ),
                              P002) +
         New_Frame_Set_Array( Window( 50 ns, 55 ns ),
                              P003) +
         New_Frame_Set_Array( Window( 50 ns, 55 ns ),
                              P012)
   )),
  --
  -- Frame Set 2
  --
 (  Delay  => Delay ( 60 ns ),
  Timing => New_Time_Data(
     New_Frame_Set_Array( Non_Return( 0 ns ), P001) +
     New_Frame_Set_Array( Non_Return( 0 ns ), P011) +
     New_Frame_Set_Array( Non_Return( 0 ns ), P010) +
     New_Frame_Set_Array( Pulse_low( 15 ns, 40 ns ), P002) +
     New_Frame_Set_Array( Non_Return( 0 ns ), P003) +
     New_Frame_Set_Array( Window( 50 ns, 55 ns ), P012)
     )),
  --
  -- Frame Set 3
  --
(  Delay  => Delay ( 60 ns ),
```

```
      Timing => New_Time_Data(
         New_Frame_Set_Array( Non_Return( 0 ns ), P001) +
         New_Frame_Set_Array( Non_Return( 0 ns ), P011) +
         New_Frame_Set_Array( Window( 50 ns, 55 ns ), P010) +
         New_Frame_Set_Array( Pulse_low( 15 ns, 40 ns ), P002) +
         New_Frame_Set_Array( Window( 50 ns, 55 ns ), P003) +
         New_Frame_Set_Array( Window( 50 ns, 55 ns ), P012)
         )),
        --
        -- Frame Set 4
        --
   (  Delay  => Delay ( 60 ns ),
      Timing => New_Time_Data(
         New_Frame_Set_Array( Non_Return( 0 ns ), P001) +
         New_Frame_Set_Array( Window( 50 ns, 55 ns ), P011) +
         New_Frame_Set_Array( Non_Return( 0 ns ), P010) +
         New_Frame_Set_Array( Pulse_low( 15 ns, 40 ns ), P002) +
         New_Frame_Set_Array( Window( 50 ns, 55 ns ), P003) +
         New_Frame_Set_Array( Window( 50 ns, 55 ns ), P012)
         ) ) );

 BEGIN
   loop
     READ_FILE_SLICE (vector_file, Vector);  -- get first vector
          exit when vector.end_of_file;
          apply(wpl, vector.codes.all, WTL(vector.fs_integer));
   end loop;
      END waveform;
END WGP_am2901;
```

Figure 9-5. Waveform Generator Procedure (WGP) for the AM2901

The WGP begins with a set of context clauses necessary to make packages visible to the WGP. Next, we need to declare an external test vector file which is read by the WGP. The name of the external file is *vectors.txt* and the contents of the file will be provided in Section 9.2.4. Next, we group the input and output pins to efficiently define the frame set array. We have defined a total of six pin groups (i.e., pinsets): P001, P011, P010, P002, P003, and P012. This pin grouping information came from the tester **Pin_Lists.H** file shown earlier in Figure 9-2. The **Pin_Lists.H** file contains the following six pin groupings:

```
pin_group_001 =  I8 I7 I6 I5 I4 I3 I2 I1 I0 A3 A2 A1 A0
                 B3 B2 B1 B0 D3 D2 D1 D0 CN OE;
pin_group_011 =  RAM3 Q3;
```

```
pin_group_010 =  RAM0 Q0;
pin_group_002 =  CP;
pin_group_003 =  Y3 Y2 Y1 Y0;
pin_group_012 =   P  G CN4 OVR  Z F3;
```

Next, we declare multiple pin code sets to handle the different operational modes of the AM2901. Since this bit-slice processor contains bi-directional pins and supports multiple operation modes, we need four timing sets (frame sets). The number of timing sets required to test this device was specified in the tester file "TB.db". Figure 9-6 shows the content of the file TB.db.

```
;written by db953: V3.12, from /tmp/teradynetemp.acig
(dbase pinmap "channel_map.h")

(dbase pins (pins-form)
(default :comp on :strobe edge :check on :mux off
:io (zstate t0) :driver-load off :mcg off)
)

(dbase ac-spec (spec-form


 :symbol      :item  :description  :units :devspec)
(tprop1      "tp1" "propagation delay" "ns" yes)
)

(dbase ac-spectable acspecs (table-form
:symbol    :min     :max    :value   :start    :stop    :from     :to)
(tprop1     0.0   500.0   500.0     0.0      0.0      0.0      0.0)
)
(dbase ac-calc
(calc-form  :t1 :t2 :fmt)
      (default :units "ns" :t3  6500.0ns)
 (set :at 0 :label w0
  :period  "60ns"
  (pin_group_001 "0ns" "57.0ns" nr0)
  (pin_group_011 "0ns" "57.0ns" nr0)
  (pin_group_010 "0ns" "57.0ns" nr0)
  (pin_group_002 "15ns" "40ns" p1)
  (pin_group_003 "50ns" "55ns" ek0)
  (pin_group_012 "50ns" "55ns" ek0)
 )

 (set :label w1
  :period  "60ns"
```

```
  (pin_group_001 "0ns" "57.0ns" nr0)
  (pin_group_011 "0ns" "57.0ns" nr0)
  (pin_group_010 "0ns" "57.0ns" nr0)
  (pin_group_002 "15ns" "40ns" pl)
  (pin_group_003 "0ns" "57.0ns" nr0)
  (pin_group_012 "50ns" "55ns" ek0)
 )

 (set :label w2
  :period  "60ns"
  (pin_group_001 "0ns" "57.0ns" nr0)
  (pin_group_011 "0ns" "57.0ns" nr0)
  (pin_group_010 "50ns" "55ns" ek0)
  (pin_group_002 "15ns" "40ns" pl)
  (pin_group_003 "50ns" "55ns" ek0)
  (pin_group_012 "50ns" "55ns" ek0)
 )

 (set :label w3
  :period  "60ns"
  (pin_group_001 "0ns" "57.0ns" nr0)
  (pin_group_011 "50ns" "55ns" ek0)
  (pin_group_010 "0ns" "57.0ns" nr0)
  (pin_group_002 "15ns" "40ns" pl)
  (pin_group_003 "50ns" "55ns" ek0)
  (pin_group_012 "50ns" "55ns" ek0)
 )
)
```

Figure 9-6. TB.db file for the AM2901 on the Teradyne J953 ATE

In addition to the number of timing sets required, this file contains all of the necessary information for defining the individual timing sets.

```
. . . .
(set :at 0 :label w0
 :period  "60ns"
  (pin_group_001 "0ns" "57.0ns" nr0)
  (pin_group_011 "0ns" "57.0ns" nr0)
  (pin_group_010 "0ns" "57.0ns" nr0)
  (pin_group_002 "15ns" "40ns" pl)
  (pin_group_003 "50ns" "55ns" ek0)
  (pin_group_012 "50ns" "55ns" ek0)
 )
. . . .
```

The segment above specifies the duration of a slice (60 ns) and associates each test pinset (pin group) with a particular ATE format. It also defines edge transitions of each ATE format. Using this information, we can create the following timing set in WAVES:

```
(Delay  => Delay ( 60 ns ),
 Timing => New_Time_Data(
  New_Frame_Set_Array( Non_Return( 0 ns ), P001) +
  New_Frame_Set_Array( Non_Return( 0 ns ), P011) +
  New_Frame_Set_Array( Non_Return( 0 ns ), P010) +
  New_Frame_Set_Array( Pulse_low( 15 ns, 40 ns ), P002) +
  New_Frame_Set_Array( Window( 50 ns, 55 ns ), P003) +
  New_Frame_Set_Array( Window( 50 ns, 55 ns ), P012)
  ))
```

As we can see, even though different syntax was used, there is a great similarity between these two representations. This translation process requires that we must be familiar with the ATE formats. For example, we need to recognize that the ATE format ***nr0*** is similar to the **Non_Return** frame set format in WAVES. The relationship between the Teradyne J953 ATE formats and the frame set formats in WAVES is shown in table form below. With this table, we can visualize this translation process more clearly.

Teradyne J953 ATE formats	Frame set formats in WAVES
nr0	Non_Return
pl	Pulse_low
ek0	Window

Finally, the WGP reads the external file, one slice of pin codes at a time, and constructs a slice of the waveform utilizing the APPLY procedure. The constructed waveform is provided to the testbench to drive the WAVES-VHDL simulation. This process repeats until the end of the external file is reached.

9.2.4 An External Test Vector File for the AM2901

We created an external test vector file which supports WAVES-VHDL simulation from the tester's vector file. The test vector file used for testing the AM2901 on the ATE Teradyne J953 is "V0006.T", and it is shown in Figure 9-7. Since this test vector file is very large, we only show a part of it. Let's examine this file briefly so that we can translate this test vector file into the external file format in WAVES.

```
****************** V0006.T **************************
#include "channel_map.h"
select suffix = ".vecig";
import label w0;
import label w1;
import label w2;
import label w3;
export vector V0006;

vector V0006
      (pin_list_5:S  pin_list_6:S  pin_list_7:S  pin_list_8:S)
{
 w w0  > 0110111110 0000000000 00000000LL LLLLHHHL;
       > 0110111110 0000000010 10000000LH LHLLHHLL;
       > 0110111110 0000000101 00000000HL HLLLHHLH;
       > 0110111110 0000000010 10000000LH LHLLHHLL;
 w w1  > 0110111110 0000000010 10000001XX XXLLHHLL;
       > 0110111110 0000000101 00000001XX XXLLHHLH;
 w w0  > 0110111110 0000000101 00000000HL HLLLHHLH;
 w w1  > 0110111110 0000000101 00000001XX XXLLHHLH;
       > 0110111110 0000000010 10000001XX XXLLHHLL;
 w w0  > 0110111110 0000000010 10000000LH LHLLHHLL;
       > 0110111110 0000000000 00000000LL LLLLHHHL;
                 . . . . . .
                    . . .
       > 0110111110 0000001000 10000000LL LHLLHHLL;
 w w2  > 1010111000 0000000000 0000L0L0LL LLLLHHHL;
       > 1010111000 0000000000 0001L0L0HL LLLLHHLH;
 rep 1 > 1010111000 0000000000 0001H0L0HH HHLHLLLH;
       > 1010111000 0000000000 0000H0L0LL LHLLHHLL;
 rep 1 > 1010111000 0000000000 0000L0L0LL LLLLHHHL;
 w w3  > 1110111000 0000000000 000L0H00LL LLLLHHHL;
       > 1110111000 0000000000 000L1H00LH HHLLHHLL;
 rep 1 > 1110111000 0000000000 000H1H00HH HHLHLLLH;
 w w0  > 0000111110 0000000000 00000000LL LLLLHHHL;
       > 0010110100 0000000000 00000000LL LLLLHHHL;
```

```
                    . . . . . . .
                        . . .
stop
        > 0011111100 0000000100 01000000HH HHHHLHLH;}
```

Figure 9-7. V0006.T for Testing the AM2901 on the Teradyne J953 ATE

First, the test vector file V0006.T explicitly specifies the ATE formats to be imported, so that they can be used in conjunction with the pin codes (import label w0) to construct waveforms for the tester. Then, it specifies the pin ordering for the test vectors. This pin ordering is specified by the statement "**Pin_List_5:S Pin_List_6:S Pin_List_7:S Pin_List_8:S**". Recall that the **Pin_Lists_5** through 7 were defined in the **Pin_Lists.H** file earlier in Figure 9-2. Concatenation of these pin lists will set the pin order for the test vector. This is similar to the fact that the pin codes in the external file must match the **Test_Pins** declaration. Incidently, the results of the concatenation produced the same pin ordering as the **Test_Pins** type shown earlier in Figure 9-3. Then, the file defines an ATE format (`w0`) to be used, and corresponding pin codes, to define a tester vector for the test cycle. From this point on, this file format is very similar to the external file format except they use different syntax. For example, we have freedom to specify only fields that are changing from the previous vector. The following demonstrates the translation process from a test vector in V0006.T to the external file format which utilizes the multiple time sets. Recall from the TB.db file, the ATE format w0 was translated to the time set 1 in the WGP. Note that, in the translation process, "L"s and "H"s in the tester file became "O"s and "1"s in the external file.

V0006.T:

```
w w0  > 0110111110 0000000000 00000000LL LLLLHHHL;
      > 0110111110 0000000010 10000000LH LHLLHHLL;
      > 0110111110 0000000101 00000000HL HLLLHHLH;
```

External file (vectors.txt):

```
0110111110 0000000000 0000000000 00001110 : 1 ;
0110111110 0000000010 1000000001 01001100 ;
0110111110 0000000101 0000000010 10001101 ;
```

In general, a typical ATE also uses various test vector compression techniques such as repeat and loop. The following example illustrates the use of the repeat command and its corresponding representation in the external file. In this example, the repeat command (`rep 1`) is translated into a simple duplication of the same vector in the external file.

V0006.T:

```
 w w2   > 1010111000 0000000000 0000L0L0LL LLLLHHHL;
        > 1010111000 0000000000 0001L0L0HL LLLLHHLH;
 rep 1  > 1010111000 0000000000 0001H0L0HH HHLHLLLH;
        > 1010111000 0000000000 0000H0L0LL LHLLHHLL;
```

External file (vectors.txt):

```
1010111000 0000000000 0000000000 00001110 : 3;
1010111000 0000000000 0001000010 00001101 ;
1010111000 0000000000 0001100011 11010001 ;
% repeat
1010111000 0000000000 0001100011 11010001 ;
1010111000 0000000000 0000100000 01001100 ;
```

Again, with some knowledge about the tester specific information, we can easily capture the ATE's test vectors in the external file for WAVES application. The external file (vectors.txt) corresponding to the V0006.T is presented in Figure 9-8.

```
0110111110 0000000000 0000000000 00001110 : 1;
0110111110 0000000010 1000000001 01001100 ;
0110111110 0000000101 0000000010 10001101 ;
0110111110 0000000010 1000000001 01001100 ;
0110111110 0000000010 10000001XX XX001100 : 2;
0110111110 0000000101 00000001XX XX001101 ;
0110111110 0000000101 0000000010 10001101 : 1;
0110111110 0000000101 00000001XX XX001101 : 2;
0110111110 0000000010 10000001XX XX001100 ;
0110111110 0000000010 1000000001 01001100 : 1;
0110111110 0000000000 0000000000 00001110 ;
                  . . . . . .
                     . . .
0110111110 0000001000 1000000000 01001100 ;
1010111000 0000000000 0000000000 00001110 : 3;
1010111000 0000000000 0001000010 00001101 ;
1010111000 0000000000 0001100011 11010001 ;
% repeat
1010111000 0000000000 0001100011 11010001 ;
1010111000 0000000000 0000100000 01001100 ;
1010111000 0000000000 0000000000 00001110 ;
% repeat
```

```
1010111000 0000000000 0000000000 00001110 ;
1110111000 0000000000 0000010000 00001110 : 4;
1110111000 0000000000 0000110001 11001100 ;
1110111000 0000000000 0001110011 11010001 ;
% repeat
1110111000 0000000000 0001110011 11010001 ;
0000111110 0000000000 0000000000 00001110 : 1;
0010110100 0000000000 0000000000 00001110 ;
                  . . . . . .
                     . . .
0011111100 0000000100 0100000011 11110101 ;
```

Figure 9-8. External Test Vector File for the AM2901

9.2.5 The Header File for the AM2901

The header file is used for documentation; we simply provide it here to complete WAVES dataset, in Figure 9-9. For more information, please refer back to Chapter 5.

```
-- ****************************************************
-- ******** Header File for Entity: AM2901
-- ****************************************************
--
-- Data Set Identification Information
--
TITLE         This WAVES dataset was created from the tester
       data of Teradyne J953 ATE to prove WAVES capability
       as a representation for exchanging information
DEVICE_ID     Bit Slice Processor

DATE          Fri Mar  13 09:25:58 1996
ORIGIN        Company X Design Team
AUTHOR        Company or Person
AUTHOR        Maybe Multiple ... Companies or People
DATE          Fri Mar  15 09:25:58 1996
ORIGIN        Modified by Company X Design Team
AUTHOR        Who did it Company or Person

OTHER         Any general comments you want
OTHER         Built Using WAVES-VHDL 1164 STD Libraries
--
-- Data Set Construction Information
--
WAVES_FILENAME   /export/home/am2901pins.vhd WORK
```

```
library            WAVES_1164;
use                WAVES_1164.WAVES_1164_Pin_Codes.all;
use                WAVES_1164.WAVES_1164_Logic_Value.all;
use                WAVES_1164.WAVES_Interface.all;
use                WORK.UUT_Test_pins.all;
WAVES_UNIT         WAVES_OBJECTS                              WORK
WAVES_FILENAME     /export/home/am2901_wgen.vhd WORK
--
EXTERNAL_FILENAME vectors.txt                          VECTORS
--
WAVEFORM_GENERATOR_PROCEDURE      WORK.waves_am2901.waveform
```

Figure 9-9. Header File for the AM2901

In ***summary of WAVES dataset for the AM2901***, we have presented the technique to capture ATE test data in a WAVES dataset. This process requires good knowledge of both tester-specific data files and WAVES. Once we have gained this knowledge, we can utilize WAVES' capability to effectively communicate or exchange information between a test and a design environment. Next, we will create a testbench that utilizes both the VHDL model of the UUT and the WAVES dataset.

9.3 A Testbench for the AM2901

Since this chapter is focused on capturing the ATE tester data using WAVES, we will not spend much time to explain the functionality of AM2901 or the quality of its test vectors. If we want to devote some time to develop the VHDL model for AM2901, we can try it. It should be a very good exercise. The device specification and functional description of the AM2901 (Bit Slice Processor, 4-bit) can be obtained from Advanced Micro Devices, Inc., or any of your local distributors. To develop this VHDL model, we provide an entity declaration for the AM2901 and an empty architecture. In addition, we have created a testbench which can be used to test the VHDL model (UUT). All we have to do, to carry out the WAVES-VHDL simulation, is create the architectural body using one of the following descriptions: behavioral, register transfer level, structural, or a combination of any. The entity and empty architecture is provided in Figure 9-10. The testbench is presented in Figure 9-11. (Enjoy your modeling.)

```
library IEEE;
use IEEE.Std_logic_1164.all;
entity am2901 is
```

```
  port ( I8 : in Std_ulogic;            I7 : in Std_ulogic;
         I6 : in Std_ulogic;            I5 : in Std_ulogic;
         I4 : in Std_ulogic;            I3 : in Std_ulogic;
         I2 : in Std_ulogic;            I1 : in Std_ulogic;
         I0 : in Std_ulogic;            A3 : in Std_ulogic;
         A2 : in Std_ulogic;            A1 : in Std_ulogic;
         A0 : in Std_ulogic;            B3 : in Std_ulogic;
         B2 : in Std_ulogic;            B1 : in Std_ulogic;
         B0 : in Std_ulogic;            D3 : in Std_ulogic;
         D2 : in Std_ulogic;            D1 : in Std_ulogic;
         D0 : in Std_ulogic;            CN : in Std_ulogic;
         CP : in Std_ulogic;            RAM3 : inout Std_logic;
         RAM0 : inout Std_logic;        Q3 : inout Std_logic;
         Q0 : inout Std_logic;          OE : in Std_ulogic;
         Y3 : out Std_ulogic;           Y2 : out Std_ulogic;
         Y1 : out Std_ulogic;           Y0 : out Std_ulogic;
          P : out Std_ulogic;           G : out Std_ulogic;
        CN4 : out Std_ulogic;           OVR : out Std_ulogic;
          Z : out Std_ulogic;           F3 : out Std_ulogic );

end am2901;

architecture empty of am2901 is
begin

 --  Add VHDL description to describe AM2901.
 --
 --  The reader might want to model only partial functionality
 --  of the device to test WAVES dataset.  In fact you can
 --  run WAVES-VHDL simulation with this empty architecture,
 -- if you are only interested in checking out WAVES
 -- dataset.  Try it.
 --
end empty;
```

Figure 9-10. Entity and Architecture Declaration for the AM2901

Our testbench, shown in Figure 9-11, consists of seven elements: context clauses, UUT component and configuration declarations, connection signal declarations, invocation of the waveform generator, translation functions, UUT component instantiation, and monitor processes. We have described the purpose of each element in detail in Section 6.3 of Chapter 6. Please refer back to Chapter 6 to review any elements of the testbench.

Here, we would like to point out a few issues to which we should pay close attention. As we can see from the entity declaration in Figure 9-10, the AM2901 has four bi-directional pins. As a result, this testbench calls the **Bi_Dir_1164** translation function to properly convert WAVES port integers into 1164 **Std_Logic** values. These function calls are shown below. (The arguments wpl.wpl(24) through wpl.wpl(27) represent the bi-directional pins in position within WAVES Port List.)

```
BI_DIREC_RAM3      <= BI_DIR_1164(wpl.wpl( 24 ));
BI_DIREC_RAM0      <= BI_DIR_1164(wpl.wpl( 25 ));
BI_DIREC_Q3        <= BI_DIR_1164(wpl.wpl( 26 ));
BI_DIREC_Q0        <= BI_DIR_1164(wpl.wpl( 27 ));
```

These converted 1164 **Std_Logic** values are assigned to intermediate signals, such as **Bi_Direc_Ram3**, and they are used to drive or observe the bi-directional pins depending on the mode of the operation. We described the translation function **Bi_Dir_1164** extensively in Section 6.2 of Chapter 6, and demonstrated its usage in Section 8.2.4 of Chapter 8. As we recall, this function was developed specifically to support the bi-directional nature of the pins.

```
LIBRARY ieee;
USE ieee.std_logic_1164.ALL;

LIBRARY waves_1164;
USE waves_1164.WAVES_1164_utilities.all;

USE WORK.UUT_test_pins.all;
USE work.waves_objects.all;

USE work.WGP_am2901.all;

ENTITY test_bench IS
END test_bench;

ARCHITECTURE am2901_test OF test_bench IS

--***********   Component Declaration    ***************

   COMPONENT am2901
    PORT ( I8  : IN   std_ulogic;          I7  : IN   std_ulogic;
           I6  : IN   std_ulogic;          I5  : IN   std_ulogic;
           I4  : IN   std_ulogic;          I3  : IN   std_ulogic;
           I2  : IN   std_ulogic;          I1  : IN   std_ulogic;
           I0  : IN   std_ulogic;          A3  : IN   std_ulogic;
           A2  : IN   std_ulogic;          A1  : IN   std_ulogic;
           A0  : IN   std_ulogic;          B3  : IN   std_ulogic;
```

```
        B2  : IN   std_ulogic;            B1  : IN   std_ulogic;
        B0  : IN   std_ulogic;            D3  : IN   std_ulogic;
        D2  : IN   std_ulogic;            D1  : IN   std_ulogic;
        D0  : IN   std_ulogic;            CN  : IN   std_ulogic;
        CP  : IN   std_ulogic;            RAM3: INOUT std_logic;
        RAM0: INOUT std_logic;            Q3  : INOUT std_logic;
        Q0  : INOUT std_logic;            OE  : IN   std_ulogic;
        Y3  : OUT  std_ulogic;            Y2  : OUT  std_ulogic;
        Y1  : OUT  std_ulogic;            Y0  : OUT  std_ulogic;
        P   : OUT  std_ulogic;            G   : OUT  std_ulogic;
        CN4 : OUT  std_ulogic;            OVR : OUT  std_ulogic;
        Z   : OUT  std_ulogic;           F3  : OUT  std_ulogic);
  END COMPONENT;

 --***********CONFIGURATION SPECIFICATION ***************

FOR ALL:am2901 USE ENTITY work.am2901(empty);

 --************************************************************
 -- stimulus signals for the waveforms mapped into UUT INPUTS
 --************************************************************

   SIGNAL WAV_STIM_I8                     :std_ulogic;
   SIGNAL WAV_STIM_I7                     :std_ulogic;
   SIGNAL WAV_STIM_I6                     :std_ulogic;
   SIGNAL WAV_STIM_I5                     :std_ulogic;
   SIGNAL WAV_STIM_I4                     :std_ulogic;
   SIGNAL WAV_STIM_I3                     :std_ulogic;
   SIGNAL WAV_STIM_I2                     :std_ulogic;
   SIGNAL WAV_STIM_I1                     :std_ulogic;
   SIGNAL WAV_STIM_I0                     :std_ulogic;
   SIGNAL WAV_STIM_A3                     :std_ulogic;
   SIGNAL WAV_STIM_A2                     :std_ulogic;
   SIGNAL WAV_STIM_A1                     :std_ulogic;
   SIGNAL WAV_STIM_A0                     :std_ulogic;
   SIGNAL WAV_STIM_B3                     :std_ulogic;
   SIGNAL WAV_STIM_B2                     :std_ulogic;
   SIGNAL WAV_STIM_B1                     :std_ulogic;
   SIGNAL WAV_STIM_B0                     :std_ulogic;
   SIGNAL WAV_STIM_D3                     :std_ulogic;
   SIGNAL WAV_STIM_D2                     :std_ulogic;
   SIGNAL WAV_STIM_D1                     :std_ulogic;
   SIGNAL WAV_STIM_D0                     :std_ulogic;
   SIGNAL WAV_STIM_CN                     :std_ulogic;
   SIGNAL WAV_STIM_CP                     :std_ulogic;
   SIGNAL WAV_STIM_OE                     :std_ulogic;
```

```
--*******************************************************
-- Expected signals used in monitoring the UUT OUTPUTS
--*******************************************************
  SIGNAL WAV_EXPECT_RAM3                          :std_logic;
  SIGNAL WAV_EXPECT_RAM0                          :std_logic;
  SIGNAL WAV_EXPECT_Q3                            :std_logic;
  SIGNAL WAV_EXPECT_Q0                            :std_logic;
  SIGNAL WAV_EXPECT_Y3                            :std_ulogic;
  SIGNAL WAV_EXPECT_Y2                            :std_ulogic;
  SIGNAL WAV_EXPECT_Y1                            :std_ulogic;
  SIGNAL WAV_EXPECT_Y0                            :std_ulogic;
  SIGNAL WAV_EXPECT_P                             :std_ulogic;
  SIGNAL WAV_EXPECT_G                             :std_ulogic;
  SIGNAL WAV_EXPECT_CN4                           :std_ulogic;
  SIGNAL WAV_EXPECT_OVR                           :std_ulogic;
  SIGNAL WAV_EXPECT_Z                             :std_ulogic;
  SIGNAL WAV_EXPECT_F3                            :std_ulogic;

--*********************************************************
-- UUT Output signals used In Monitoring ACTUAL Values
--*********************************************************

  SIGNAL ACTUAL_Y3                            :std_ulogic;
  SIGNAL ACTUAL_Y2                            :std_ulogic;
  SIGNAL ACTUAL_Y1                            :std_ulogic;
  SIGNAL ACTUAL_Y0                            :std_ulogic;
  SIGNAL ACTUAL_P                             :std_ulogic;
  SIGNAL ACTUAL_G                             :std_ulogic;
  SIGNAL ACTUAL_CN4                           :std_ulogic;
  SIGNAL ACTUAL_OVR                           :std_ulogic;
  SIGNAL ACTUAL_Z                             :std_ulogic;
  SIGNAL ACTUAL_F3                            :std_ulogic;

--************************************************************
-- Bi_directional signals used  for stimulus signals mapped
-- into UUT INPUTS and also monitoring the UUT OUTPUTS
--************************************************************

  SIGNAL BI_DIREC_RAM3                            :std_logic;
  SIGNAL BI_DIREC_RAM0                            :std_logic;
  SIGNAL BI_DIREC_Q3                              :std_logic;
  SIGNAL BI_DIREC_Q0                              :std_logic;

--
```

```
****************************************************************
  -- WAVES signals OUTPUTing each slice of the waves port list
  --
****************************************************************

       SIGNAL wpl  : WAVES_port_list;

BEGIN
  --
  --
****************************************************************
  -- process that generates WAVES waveform
  --
****************************************************************

      WAVES: waveform(wpl);

  --
****************************************************************
  -- processes that convert the WPL values to 1164 Logic Values
  --
****************************************************************

  WAV_STIM_I8                  <= STIM_1164(wpl.wpl( 1 ));
  WAV_STIM_I7                  <= STIM_1164(wpl.wpl( 2 ));
  WAV_STIM_I6                  <= STIM_1164(wpl.wpl( 3 ));
  WAV_STIM_I5                  <= STIM_1164(wpl.wpl( 4 ));
  WAV_STIM_I4                  <= STIM_1164(wpl.wpl( 5 ));
  WAV_STIM_I3                  <= STIM_1164(wpl.wpl( 6 ));
  WAV_STIM_I2                  <= STIM_1164(wpl.wpl( 7 ));
  WAV_STIM_I1                  <= STIM_1164(wpl.wpl( 8 ));
  WAV_STIM_I0                  <= STIM_1164(wpl.wpl( 9 ));
  WAV_STIM_A3                  <= STIM_1164(wpl.wpl( 10 ));
  WAV_STIM_A2                  <= STIM_1164(wpl.wpl( 11 ));
  WAV_STIM_A1                  <= STIM_1164(wpl.wpl( 12 ));
  WAV_STIM_A0                  <= STIM_1164(wpl.wpl( 13 ));
  WAV_STIM_B3                  <= STIM_1164(wpl.wpl( 14 ));
  WAV_STIM_B2                  <= STIM_1164(wpl.wpl( 15 ));
  WAV_STIM_B1                  <= STIM_1164(wpl.wpl( 16 ));
  WAV_STIM_B0                  <= STIM_1164(wpl.wpl( 17 ));
  WAV_STIM_D3                  <= STIM_1164(wpl.wpl( 18 ));
  WAV_STIM_D2                  <= STIM_1164(wpl.wpl( 19 ));
  WAV_STIM_D1                  <= STIM_1164(wpl.wpl( 20 ));
  WAV_STIM_D0                  <= STIM_1164(wpl.wpl( 21 ));
  WAV_STIM_CN                  <= STIM_1164(wpl.wpl( 22 ));
  WAV_STIM_CP                  <= STIM_1164(wpl.wpl( 23 ));
  BI_DIREC_RAM3                <= BI_DIR_1164(wpl.wpl( 24 ));
```

```
WAV_EXPECT_RAM3              <= EXPECT_1164(wpl.wpl( 24 ));
BI_DIREC_RAM0                <= BI_DIR_1164(wpl.wpl( 25 ));
WAV_EXPECT_RAM0              <= EXPECT_1164(wpl.wpl( 25 ));
BI_DIREC_Q3                  <= BI_DIR_1164(wpl.wpl( 26 ));
WAV_EXPECT_Q3                <= EXPECT_1164(wpl.wpl( 26 ));
BI_DIREC_Q0                  <= BI_DIR_1164(wpl.wpl( 27 ));
WAV_EXPECT_Q0                <= EXPECT_1164(wpl.wpl( 27 ));
WAV_STIM_OE                  <= STIM_1164(wpl.wpl( 28 ));
WAV_EXPECT_Y3                <= EXPECT_1164(wpl.wpl( 29 ));
WAV_EXPECT_Y2                <= EXPECT_1164(wpl.wpl( 30 ));
WAV_EXPECT_Y1                <= EXPECT_1164(wpl.wpl( 31 ));
WAV_EXPECT_Y0                <= EXPECT_1164(wpl.wpl( 32 ));
WAV_EXPECT_P                 <= EXPECT_1164(wpl.wpl( 33 ));
WAV_EXPECT_G                 <= EXPECT_1164(wpl.wpl( 34 ));
WAV_EXPECT_CN4               <= EXPECT_1164(wpl.wpl( 35 ));
WAV_EXPECT_OVR               <= EXPECT_1164(wpl.wpl( 36 ));
WAV_EXPECT_Z                 <= EXPECT_1164(wpl.wpl( 37 ));
WAV_EXPECT_F3                <= EXPECT_1164(wpl.wpl( 38 ));

--****************************************
-- UUT Port Map - Name Symantics Denote Usage
--****************************************

u1: am2901
PORT MAP(
  I8   => WAV_STIM_I8,     I7   => WAV_STIM_I7,
  I6   => WAV_STIM_I6,     I5   => WAV_STIM_I5,
  I4   => WAV_STIM_I4,     I3   => WAV_STIM_I3,
  I2   => WAV_STIM_I2,     I1   => WAV_STIM_I1,
  I0   => WAV_STIM_I0,     A3   => WAV_STIM_A3,
  A2   => WAV_STIM_A2,     A1   => WAV_STIM_A1,
  A0   => WAV_STIM_A0,     B3   => WAV_STIM_B3,
  B2   => WAV_STIM_B2,     B1   => WAV_STIM_B1,
  B0   => WAV_STIM_B0,     D3   => WAV_STIM_D3,
  D2   => WAV_STIM_D2,     D1   => WAV_STIM_D1,
  D0   => WAV_STIM_D0,     CN   => WAV_STIM_CN,
  CP   => WAV_STIM_CP,     RAM3=> BI_DIREC_RAM3,
  RAM0=> BI_DIREC_RAM0,    Q3   => BI_DIREC_Q3,
  Q0   => BI_DIREC_Q0,     OE   => WAV_STIM_OE,
  Y3   => ACTUAL_Y3,       Y2   => ACTUAL_Y2,
  Y1   => ACTUAL_Y1,       Y0   => ACTUAL_Y0,
  P    => ACTUAL_P,        G    => ACTUAL_G,
  CN4  => ACTUAL_CN4,      OVR  => ACTUAL_OVR,
  Z    => ACTUAL_Z,        F3   => ACTUAL_F3);
--*******************************************************
-- Monitor Processes To Verify The UUT Operational Response
--*******************************************************
```

```
Monitor_RAM3:
  PROCESS(BI_DIREC_RAM3, WAV_expect_RAM3)
  BEGIN
       assert(Compatible (actual => BI_DIREC_RAM3,
                          expected => WAV_expect_RAM3))
       report "Error on RAM3 output" severity WARNING;
  END PROCESS;


Monitor_RAM0:
  PROCESS(BI_DIREC_RAM0, WAV_expect_RAM0)
  BEGIN
       assert(Compatible (actual => BI_DIREC_RAM0,
                          expected => WAV_expect_RAM0))
       report "Error on RAM0 output" severity WARNING;
  END PROCESS;


Monitor_Q3:
  PROCESS(BI_DIREC_Q3, WAV_expect_Q3)
  BEGIN
       assert(Compatible (actual => BI_DIREC_Q3,
                          expected => WAV_expect_Q3))
       report "Error on Q3 output" severity WARNING;
  END PROCESS;


Monitor_Q0:
  PROCESS(BI_DIREC_Q0, WAV_expect_Q0)
  BEGIN
       assert(Compatible (actual => BI_DIREC_Q0,
                          expected => WAV_expect_Q0))
       report "Error on Q0 output" severity WARNING;
  END PROCESS;


Monitor_Y3:
  PROCESS(ACTUAL_Y3, WAV_expect_Y3)
  BEGIN
       assert(Compatible (actual => ACTUAL_Y3,
                          expected => WAV_expect_Y3))
       report "Error on Y3 output" severity WARNING;
  END PROCESS;


Monitor_Y2:
```

```
   PROCESS(ACTUAL_Y2, WAV_expect_Y2)
   BEGIN
        assert(Compatible (actual => ACTUAL_Y2,
                          expected => WAV_expect_Y2))
        report "Error on Y2 output" severity WARNING;
   END PROCESS;

 Monitor_Y1:
   PROCESS(ACTUAL_Y1, WAV_expect_Y1)
   BEGIN
        assert(Compatible (actual => ACTUAL_Y1,
                          expected => WAV_expect_Y1))
        report "Error on Y1 output" severity WARNING;
   END PROCESS;

 Monitor_Y0:
   PROCESS(ACTUAL_Y0, WAV_expect_Y0)
   BEGIN
        assert(Compatible (actual => ACTUAL_Y0,
                          expected => WAV_expect_Y0))
        report "Error on Y0 output" severity WARNING;
   END PROCESS;

 Monitor_P:
   PROCESS(ACTUAL_P, WAV_expect_P)
   BEGIN
        assert(Compatible (actual => ACTUAL_P,
                          expected => WAV_expect_P))
        report "Error on P output" severity WARNING;
   END PROCESS;

 Monitor_G:
   PROCESS(ACTUAL_G, WAV_expect_G)
   BEGIN
        assert(Compatible (actual => ACTUAL_G,
                          expected => WAV_expect_G))
        report "Error on G output" severity WARNING;
   END PROCESS;
 Monitor_CN4:
   PROCESS(ACTUAL_CN4, WAV_expect_CN4)
   BEGIN

        assert(Compatible (actual => ACTUAL_CN4,
                          expected => WAV_expect_CN4))
        report "Error on CN4 output" severity WARNING;
   END PROCESS;
```

```
 Monitor_OVR:
   PROCESS(ACTUAL_OVR, WAV_expect_OVR)
   BEGIN
        assert(Compatible (actual => ACTUAL_OVR,
                          expected => WAV_expect_OVR))
        report "Error on OVR output" severity WARNING;
   END PROCESS;


 Monitor_Z:
   PROCESS(ACTUAL_Z, WAV_expect_Z)
   BEGIN
        assert(Compatible (actual => ACTUAL_Z,
                          expected => WAV_expect_Z))
        report "Error on Z output" severity WARNING;
   END PROCESS;


 Monitor_F3:
   PROCESS(ACTUAL_F3, WAV_expect_F3)
   BEGIN
        assert(Compatible (actual => ACTUAL_F3,
                          expected => WAV_expect_F3))
        report "Error on F3 output" severity WARNING;
   END PROCESS;


END am2901_test;
```

Figure 9-11. Testbench for the AM2901

In ***summary of our testbench for the AM2901,*** we provided an entity-architecture pair and a testbench to assist us in developing the VHDL model for the AM2901. We also introduced the seven basic elements that make up a typical testbench. In particular, we have utilized the translation function **Bi_Dir_1164** in the testbench to support the bi-directional pin devices. Here, we demonstrated that the testbench generation process is systematic and only minor changes to the testbench are required to support complex devices.

In ***summary of Chapter 9***, we have described WAVES' capability as a standard for exchanging information between a design and a test environment. In particular, we demonstrated a way WAVES can be used to represent ATE test vectors. Even though additional knowledge about the ATE tester data was required, we showed that WAVES can be a very useful tool for exchanging information.

Based on the material in this chapter, we conclude that WAVES enables us to exchange test information across the application domain and organizations. Now it is time to extend WAVES further. So far, all of our examples utilized WAVES Level 1 constructs. Recall that the main purpose of WAVES Level 1 was to support the ATE environment. However, there are some situations where it is more efficient to describe a test algorithmically instead of as a flattened, bit-vector format (level 1). WAVES Level 2 can support this algorithmic test description. Level 2 of WAVES allows us to describe a test with a great deal of flexibility, and it is the topic of the next chapter.

CHAPTER 10. WAVES Level 2

Some additional capability and flexibility

In this chapter in the Advanced Topics portion of our book, we set WAVES into a more proper context. In doing so, we discover some advantages in additional capability and flexibility, and that the application of these advantages is subject to some constraints and limitations.

To begin, the WAVES standard is defined in terms of *two subsets* of VHDL: WAVES Level 1 and WAVES Level 2. We have discussed WAVES Level 1 in the previous chapters of this book. WAVES Level 1 is a proper subset of WAVES Level 2 which in turn is a proper subset of VHDL constructs.

WAVES Level 2 provides the full power of a procedural programming language for the specification of waveform data. Waveform generation procedures in WAVES Level 1 are akin to regular expressions. Thus, WAVES Level 2 is *significantly more expressive* than WAVES Level 1. In addition, WAVES Level 2 permits user-defined external data formats by removing all restrictions on the external file format.

In this chapter, we discuss WAVES Level 2 syntax and semantics in contrast with Level 1 and VHDL. We begin with a discussion of how external files are handled in Level 2. Next, we discuss the Level 2 aspects of waveform generation procedures. Then we describe, and give some examples of, level 2 constructs. Finally, we conclude our chapter with a description and example of level 2 usage. We now begin with the external file issues.

10.1 External Files in Level 2

As we discussed in Chapter 7, the external files in WAVES datasets conforming to Level 1 must follow a format defined by the WAVES standard. however, external files in Level 2 datasets need not follow this format. Hence, the external data can be specified in several ways in Level 2 datasets. We summarize these differences below, in the context of the external file formats.

1. WAVES Standard External File Format - in level 2, the external file can be read with or without the standard **Read_File_Slice** procedure, as we explain below:

(a) *With* the standard **Read_File_Slice** procedure

WAVES Level 1 datasets are also WAVES Level 2 datasets. Hence, external data files conforming to the WAVES external file format can be used in Level 2 datasets. In addition, the **Read_File_Slice** procedure defined in the **Waves_Objects** package can be used to read these external files and convert that data into standard WAVES objects. In this case, the interpretation of the external data is exactly the same as in Level 1.

(b) *Without* the standard **Read_File_Slice** procedure

WAVES Level 2 does not require the use of the standard **Read_File_Slice** procedure defined in the WAVES standard packages. Users can write their own external file readers, in the subset of VHDL permitted by WAVES Level 2. In this case, while the syntax of the external file complies with the WAVES external file format requirements, the semantics does not. The meaning of the external data is determined by the user-defined reader of the external files.

2. User-Defined External File Formats - In Level 2, the user may define the external file format, permitting additional flexibility in use.

Level 2 does not require that the external files be in the format defined by the WAVES Level 1 standard. Hence, users are free to use any VHDL-readable file formats. In this case, the users should write procedures to read the external files, properly interpret the external data and convert this data into standard WAVES objects. Thus the meaning of the data in the external files is defined solely by the user-defined file reader procedures. This opens the possibility of using abstract data representations, for example - assembly language programs for processor designs, in the external files.

We see, then, that there is considerably more flexibility permitted in level 2, for reading the external files. however, such flexibility does not extend to writing these files. WAVES Level 2 restricts the procedures which read external files to reference only the following objects from the VHDL **Std.TextIO** package: the types *Line* and *Text*; the procedures *ReadLine* and *Read*; and the function *EndFile*. These restrictions imply that *writing* external files is not permitted in WAVES Level 2 datasets; only reading is allowed.

10.2 Waveform Generation Procedures in Level 2

WAVES Level 2 permits *parametric* waveform specifications by allowing additional parameters to specify the waveform generation procedures, beyond the required **Waves_Port_List** and the optional **Waves_Match_List**. Thus, both match mode and non-match mode waveforms can be parameterized. Level 1 permits no such additional parameters; waveforms generated by Level 1 procedures are *nonparametric* waveforms.

These additional parameters which apply to the Level 2 waveform generator procedures must have the class *constant.* Since many programming constructs are permitted in Level 2, we may use these parameters to influence the generation of the waveform in interesting ways.

10.3 WAVES Level 2 Constructs

In terms of constructs, WAVES Level 2 files are legal VHDL files. Therefore, all reserved words in VHDL are also reserved in WAVES, whether they are used in WAVES syntax or not. Hence, WAVES Level 2 retains all the restrictions imposed on VHDL by WAVES Level 1, except that the following restrictions are removed:

1. Variable Assignment Statement

The most powerful feature restored in Level 2 is the variable assignment statement. This makes Level 2 tools behave similar to compilers or interpreters, whereas Level 1 tools behave as expression evaluators.

2. Type Definitions

The following type definitions are restored: access type, floating type, composite types, incomplete types, and subtypes. In addition, type conversion is permitted, and the allocator construct is restored. Attribute specifications are permitted. Character literals are restored to enumeration literals and scalar types can be of integer type. Variables can be any of these types and, together with the variable assignment statement and other control flow constructs permitted, bring the full power of a sequential programming language to WAVES Level 2.

Essentially, a WAVES Level 2 program amounts to a collection of packages within which types and subprograms are defined. Some of these subprograms are waveform generator procedures that can be invoked by concurrent procedure calls in a VHDL test-bench.

WAVES Level 2 retains all other restrictions imposed by Level 1 on VHDL. These include the complete elimination of concurrent statements, design encapsulation statements, design configuration management statements, signal assignment statement, wait statements, physical types, operator overloading and subprogram overloading.

In the following subsections, we elaborate further upon the constructs in WAVES Level 2, in comparison with VHDL and Level 1.

10.3.1 VHDL Constructs Removed in WAVES Level 2

The VHDL constructs not present in WAVES are primarily used to represent the concurrent structure of digital systems, and the parallelism inherent in these systems. WAVES Level 2 does not have the following constructs:

- **Constructs to represent design organization and structure:** Since a WAVES specification is not meant to describe the structure of the design under test, or the testbench used to exercise this design, the following constructs are unnecessary:

  Entity Specifications: These are only used to define the interface between a design object and its environment. Since WAVES datasets have no design objects, these are unnecessary.

  Architecture Specifications: These define the body of a design entity. Since WAVES does not specify entities, these are unnecessary.

  Component Specifications: A design entity may include other design entities; these are specified as *components*. Again, WAVES has no design entities, so components are unnecessary.

  Configuration Specifications: These bind specific component specifications of a design entity to specific architectures of that entity. Once again, WAVES has no entities, so configurations are unnecessary.

- **Constructs to represent concurrency:** WAVES is a sequential programming language used to determine and specify the stimulus and expected response for a design under test. Thus, the following are not needed:

Signal Declarations & Assignments: A signal is a design object with a past history of values. Since WAVES is purely sequential, the history of an object is irrelevant; thus, signals (both declarations and assignments) are unnecessary. The only exception to this involves the interaction of the WAVES dataset with the environment. This interaction takes place by means of the **Waves_Port_List** and **Waves_Match_List** signals. We have discussed the usage of this interface in detail in the previous chapters.

Wait Statements: A wait statement suspends the execution of a process. Since processes are inherently concurrent, they are not implemented in WAVES. Thus, wait statements are unnecessary in WAVES.

Disconnection Specifications: These are used in VHDL to explicitly schedule the disconnection of the driver of a guarded signal. Since WAVES does not support signals, there is no need for it to support these, either.

- **Other constructs to simplify specifications:** A few other constructs which have meaning in VHDL are not included in WAVES Level 2. These include:

  Aliases: An alias is an alternative name for referencing an object.

  Attributes: An attribute is a method of associating some additional information with a design object. Again, most VHDL design objects are not present in WAVES, so attributes were removed as well.

  File Types: Specifically, *most* file types are not supported in WAVES; the only file type supported by WAVES is *TEXT*. Other file types (declared using *file of X*) are not allowed.

  Physical Types: Physical types are used to represent measurements of some quantity. However, WAVES only represents waveforms and test vectors, neither of which have any measurable quantities; thus, physical types have been removed from WAVES.

In addition, there are a number of other restrictions listed in the WAVES reference manual. Most of these refer to the visibility of library packages and defined data types.

In summary, valid WAVES constructs include procedures, functions, user-defined data types, variable declarations, and other sequential programming constructs. This is sufficient to specify the waveform generation procedure, which is the heart of a WAVES specification.

10.3.2 WAVES Level 2 Added Constructs, not present in WAVES Level 1

As we mentioned above, WAVES Level 1 is a subset of WAVES Level 2. Level 1 is sufficient to specify most waveforms as lists of stimulii and expected responses. However, this is insufficient to specify test sets for more complex systems, and more constructs are necessary. These constructs, which we enumerate below, are included in WAVES Level 2.

- **Expanded Type System:** WAVES Level 1 only has a basic type system. WAVES Level 2 also allows:

  Floating Types: Floating point numbers are available.

  Composite Data Types: Both arrays of items and records of related items are available.

  Indirect Access Types: VHDL has operators (*allocators*) which generate objects having no name (i.e., they cannot be directly addressed), but can be referred to using *access types*. In a general programming language, this is similar to a pointer to an object. An additional type specification called an *incomplete type* is needed to resolve problems with recursive access types. These constructs are all available in WAVES Level 2.

  Type Conversions: Given two different types, VHDL has various type conversion functions. These are also available in WAVES Level 2.

  Subtypes: Given a type definition, subtypes are also available.

- **Variable Assignments:** Variables in a WAVES Level 1 dataset must be defined at declaration time, and cannot be assigned a different value at any other time. WAVES Level 2 does allow a variable assignment statement; this allows variables defined in procedures and functions to be assigned and re-assigned as needed.

- **Improved File I/O:** WAVES Level 2 is allowed to access a few procedures and functions from *Std.TextIO* regarding file input/output. In Level 1, the only file I/O procedure available is the WAVES procedure **Read_File_Slice**.

Level 2 adds some of the normal file I/O procedures from VHDL, including *Read* and *ReadLine*, but writing files is not permitted.

From these constructs, we can clearly see some of the advantages of WAVES Level 2 over WAVES Level 1:

- An improved type system allows for more flexibility in user-defined data structures.
- Variable assignment statements allow for more flexibility in user-defined procedures and functions.
- Improved file I/O procedures allow for different types of external files to be read in and interpreted by the waveform generator procedure.

We discuss each of these construct advantages further, and present some examples, in the following section.

10.3.3 WAVES Level 2 Construct Examples

A WAVES Level 2 testbench will function similarly to a WAVES Level 1 testbench, except that the user-defined functions, including the waveform generator procedure, may be more complex, and the external files may have more flexible data formats. We discuss the extra features of WAVES Level 2 which allow this in this section. Designers familiar with WAVES Level 1 can use this section as an introduction to the new features which are available in WAVES Level 2. Designers familiar with VHDL may wish to skip this section.

10.3.3.1 Data Types in WAVES Level 2

Here, we discuss and give examples of base data types, composite data types, and access types applicable to the constructs advantages in Level 2 we introduced earlier.

Base Data Types: WAVES Level 2 allows three different base data types: enumeration types, integer types, floating point types. We describe all these individually, below:

Enumeration Types: Given a set of possible values for a variable, an enumerated type simply lists all possible values:

```
type LOGIC_LEVEL is (LOW,HIGH,RISING,FALLING,UNKNOWN);
```

```
type BIT is ('0', '1');
```

These enumerated values can be used in more than one data type; the type of a value can be determined from its context:

```
type SWITCH_LEVEL is ('0','1','X');
```

Integer Types: An integer type can be declared by specifying the range of possible values for the type:

```
type BYTE_RANGE is range 0 to 255;
type WORD_INDEX is range 31 downto 0;
```

Note that the ranges are specified as a "left" endpoint and a "right" endpoint, and the words `to` or `downto` are used accordingly.

There is also the predefined integer type `INTEGER`, which is guaranteed to include -2147483647 to 2147483647. Due to the implementation-dependent methods of representing integer values, a user-defined integer range cannot contain numbers outside these bounds.

Floating Point Types: While WAVES Level 1 only allows enumerated and integer data types (above), WAVES Level 2 also allows floating point types. These are specified exactly the same way as integer types:

```
type VALID_AREA is range 0 to 1.0E06;
```

There is also the predefined floating point type `REAL`, which is guaranteed to include -1.0E38 to 1.0E38.

Composite Data Types: Also allowed in WAVES Level 2 are two types of composite data types, arrays and records:

Arrays: An array consists of a number of elements that have the same subtype. Each array element can be referenced by the name of the array and one (or more) indices that specify which element. An array type is specified by an integer range and an element subtype:

```
type WORD is array (15 downto 0) of BIT;
type KMEM is array (0 to 1023, 15 downto 0) of BIT;
```

In the case of KMEM, above, the type is a two-dimensional array, as noted by the two index ranges. Alternately, an array's subtype could be an array as well:

```
type KMEM2 is array (0 to 1023) of WORD;
```

Additionally, the index range can be left unspecified at the type declaration, and specified at each variable declaration:

```
type MEMORY is array (INTEGER range <>) of WORD;

variable MEM1 : MEMORY(0 to 65535);
variable MEM2 : MEMORY(1023 downto 0);
```

In this example, the type MEMORY is defined to be an array of an unspecified number of WORDs. MEM1 is then declared as an array of 65536 WORDs. MEM2 is declared as an array of 1024 WORDs, numbered by decreasing index.

Predefined array types include `STRING` and `BIT_VECTOR` as defined in VHDL.

Records: A record consists of a number of elements which may be of different types. Each element (*field*) has its own name, which must be distinct:

```
type DATE is
  record
    DAY   : INTEGER range 1 to 31;
    MONTH : INTEGER range 1 to 12;
    YEAR  : INTEGER range 0 to 4000;
  end record;

variable today : DATE;
```

The variable `today` then consists of three fields, `today.DAY`, `today.MONTH`, and `today.YEAR`.

Access Types: WAVES Level 2 also allows the use of allocators, de-allocators, and access types, which allow indirect referencing of objects in VHDL (similar to the use of pointers in most high-level programming languages).

Regular design objects created by some object declaration statement are given specific names by the declarations. However, the use of an *allocator* creates an object which does not have a specific name; this object is referenced using an *access value*. Access type declarations specify the type of access values. For example:

```
type ADDRESS is access MEMORY;

variable INST_ADDR : ADDRESS;
```

In this case, the access type ADDRESS refers to an object of type MEMORY. The access variable INST_ADDR then refers to some object of type MEMORY. INST_ADDR can then "point" to any MEMORY object visible at that time.

Objects created by allocators are similarly destroyed by *de-allocators*, and the use of access types to "point" to records containing other access types may cause self-referential type definitions that can only be implemented using *incomplete type declarations*. For more information about access types, allocators and de-allocators, and incomplete types, we should consult the VHDL LRM.

10.3.3.2 Subprograms in WAVES Level 2

Both WAVES Level 1 and WAVES Level 2 allow the use of subprograms: procedures and functions. These subprograms are specified exactly as in VHDL. However, one major difference regards the use of subprogram variables. WAVES Level 1 *does not* allow a variable to be re-assigned during evaluation of a subprogram; it must be given a value when it is declared, and this value cannot change. However, WAVES Level 2 *does* allow variable assignment statements to appear in a subprogram. Variable declaration statements are of the form:

```
variable <<name>> : <<type>> [ :=<<expression>> ] ;
```

This creates a variable *name* of type *type*; if an initial value is specified, the *expression* is evaluated and its value is used as the initial value. If there is no initial value, its value is taken to be *type'Left* (that is, the "first" value in *type*) if *type* is a scalar, or a corresponding set of *subtype'Left* if *type* refers to a composite data type.

Variable assignment statements take the form:

```
<<name>> := <<expression>> ;
```

This causes *expression* to be evaluated, and the result is stored in *name*.

Beyond this difference between WAVES Level 1 and WAVES Level 2, both levels allow all sequential programming constructs which are present in VHDL.

These include loop constructs, if/elsif/else, case, and so on. For more information about subprograms, we may consult any VHDL reference, such as the VHDL LRM.

10.3.3.3 File I/O in WAVES Level 2

VHDL libraries include the file I/O package *Std.TextIO*. While the majority of this package is not visible in a WAVES dataset, there are some package subprograms which can be used to read external files.

In WAVES Level 1, the only package element visible to a WAVES subprogram is the type *Text*, which is used to define the type of the external files. The remaining file I/O functions are not allowed. Instead, the subprogram **Read_File_Slice** is defined in the **Waves_Objects** library. This allows an external file to be envisioned as a list of slice specifications. A typical Level 1 waveform generator simply reads in a file slice (as a list of pin codes), interprets the pin codes, and applies the resulting slice to the device under test (where *Match* subprograms are used to check for output discrepancies).

In WAVES Level 2, however, the file I/O subprograms regarding input of text files allowed are:

- The type *Line*
- The type *Text*
- The procedure *ReadLine*
- The overloaded procedures *Read*
- The function *EndFile*

Similar subprograms regarding input of non-text files or output of any files are not available. Thus, WAVES Level 2 has the required subprograms to read text files, but cannot do any other file I/O.

However, even this limited file I/O is sufficient for a WAVES Level 2 testbench to read an external file. Other subprograms can be defined to interpret the contents and call the relevant WAVES subprograms. Thus, WAVES Level 2 does allow a more flexible external file format than WAVES Level 1. However, the bulk of the file interpretation is the *designer's responsibility*.

10.4 WAVES Level 2 Usage

To illustrate how the features available in WAVES Level 2 can be used, we will discuss an example. Our illustrative example is concerned with testing a small calculator. The VHDL model of the calculator is shown on the next page:

```
use work.constants.all;

entity calculator is
  port(A: in integer; B: in integer; opcode: in op_type; C: out
       integer);
  end calculator;

architecture behav of calculator is

begin
  CALC: process(A,B,opcode)
    variable accumulator: integer := 0;
    variable memory: integer := 0;

  begin
    case opcode is
      when add =>
         accumulator := A + B;
      when adda =>
         accumulator := accumulator + A;
      when sub =>
         accumulator := A - B;
      when suba =>
         accumulator := accumulator - A;
      when div =>
         accumulator := A / B;
      when diva =>
         accumulator := accumulator / A;
      when mul =>
         accumulator := A * B;
      when mula =>
         accumulator := accumulator * A;
      when nega =>
         accumulator := -A;
      when negacc =>
         accumulator := -accumulator;
      when clear =>
         accumulator := 0;
      when str =>
         memory := accumulator;
      when rstr =>
         accumulator := memory;
    end case;
    C<= accumulator after 1 NS;
  end process CALC;
end Behav;
```

The calculator accepts two integers and an operation code. It performs an accumulator operation based on the opcode and generates a result on the output port after a delay of 1 ns. **Op_type** is declared in the package *Constants* which is included through the *use* clause in the above model. The following is the content of the *Constants* package:

```
package Constants is

type op_type is (add, adda, sub, suba, div, diva, mul, mula,
                 nega, negacc, clear, str, rstr);
constant input_line_length : natural:= 64;
constant max_op_code_len : natural := 6;
end Constants;
```

We wish to write a WAVES test-bench to test the calculator. Our external files contain a sequence of calculator commands. An example external file **Calc_com.txt** is shown below:

```
CLEAR
SUB 76 75
SUBA 10
ADD 55 78
ADDA 20
DIV 30060 21
DIVA 2
MUL 14 34
MULA 17
STR
NEGA -45
RSTR
NEGACC
CLEAR
```

File **Calc_ans.txt** contains the expected results. The expected results for the above sequence of commands are shown below:

```
0
1
-9
133
153
1431
```

```
715
476
8092
8092
45
8092
-8092
0
```

Our WAVES Level 2 waveform generation procedure for the calculator model will read the command sequence from the file **Calccom.txt** and prepare the stimuli waveforms. The stimuli are prepared using IEEE standard logic conventions. The VHDL test bench will perform the necessary conversion between the standard logic type and the integer type used by the entity interface of the calculator model.

We will begin by showing the test pins declaration:

```
PACKAGE uut_test_pins is

TYPE test_pins is (a15, a14, a13, a12, a11, a10, a9, a8, a7, a6,
a5, a4, a3, a2, a1, a0, b15, b14, b13, b12, b11, b10, b9, b8,
b7, b6, b5, b4, b3, b2, b1, b0, op_code3, op_code2, op_code1,
op_code0, c31, c30, c29, c28, c27, c26, c25, c24, c23, c22, c21,
c20, c19, c18, c17, c16, c15, c14, c13, c12, c11, c10, c9, c8,
c7, c6, c5, c4, c3, c2, c1, c0);

END uut_test_pins;
```

Our intention is to generate waveforms, assuming that the two input integers of the calculator can be represented by sixteen bits each of the standard logic type and the output can be represented by thirty-two bits. Likewise, the opcode will be represented by four bits.

We will first define a number of functions that will be used by the waveform generator. We will begin with the procedure `int_to_pin_codes`, used to convert integers into pin codes using the IEEE standard logic representation:

```
  PROCEDURE int_to_pin_codes(variable num : in integer;
                             variable pc_string : out string;
                             constant len : in integer) is
    variable tmp : integer;
    variable orig : integer;
    variable cnt : integer;
    variable tmp_pc_string : string(pc_string'range);
```

```
    variable iline : line;

  BEGIN
    tmp := num;
    if (len < 32) then
       assert ((tmp <  (2**(len-1)-1)) and (tmp > -(2**(len-1)-
1))) report "Integer to large for two's complement vector"
         severity warning;
    end if;

    orig := num;
    for I in 1 to len loop
      if (tmp mod 2 = 1) then tmp_pc_string(len - i + 1) := '1';
      else tmp_pc_string(len - I + 1) := '0';
      end if;
      tmp := tmp / 2;
    end loop;

    if (orig<0) then
      for i in 1 to (len) loop
        if (tmp_pc_string(i) = '0') then
          tmp_pc_string(i) := '1';
        else tmp_pc_string(i) := '0';
        end if;
      end loop;

      cnt ;= len;

      loop
         if (tmp_pc_string(cnt) = '0') then
           tmp_pc_string(cnt) := '1';
         else tmp_pc_string(cnt) := '0';
         end if;
         exit when ((tmp_pc_string(cnt) = '1') or (cnt = 1));
         cnt := cnt - 1;
      end loop;
    end if;

    pc_string := tmp_pc_string;
  End int_to_pin_codes;
```

Function `strcmp` compares two strings:

```
  function strcmp( str1 : in string ; str2 : in string ) return
      boolean is
    variable str1_ptr : natural;
    variable str2_ptr : natural;
```

```
begin
  if ( str1'length /= str2'length ) then
    return false;
  end if;
  for ctr in 1 to str1'length loop
    str1_ptr := str1'left - 1 + ctr;
    str2_ptr := str2'left - 1 + ctr;
    if ( str1( str1_ptr ) /= str2( str2_ptr ) ) then
      return false;
    end if;
  end loop;

  return true;
end strcmp;
```

Procedure **skip_ws** is used to skip white space (tabs and spaces) and comments while reading the external files:

```
procedure skip_ws( str : in string;
                   ptr : out natural;
                   comment : out boolean ) is
  variable done : boolean := false;
  variable tmp_ptr : natural;
begin
  tmp_ptr := str'left;
  comment := false;
  while ( done = false ) loop
    case str(tmp_ptr) is
      when ' ' => tmp_ptr := tmp_ptr + 1;-- space
      when ht => tmp_ptr := tmp_ptr + 1;-- ht: horizontal tab
      when ff => tmp_ptr := tmp_ptr + 1;-- ff: form feed
      when '#' => comment := true;   -- #:  comment statement
                  done := true;
      when others => done := true;
    end case;
    if ( tmp_ptr > str'right ) then
      done := true;
    end if;
  end loop;
  ptr := tmp_ptr;
end skip_ws;
```

Procedure **Read_opcode** is used to read the opcode from the external command file:

```
procedure Read_opcode (variable iline: inout line;
```

```
                        variable opcode : out op_type) is
variable strg : string(1 to max_op_code_len + 1);
variable comment : boolean := false;
variable done : boolean := false;
variable cur_char : character;
variable cnt : natural := 1;

BEGIN
  for i in strg'range loop
    strg(i) := ' ';
  end loop;
  while (done = false) loop -- skipping white space
    if (iline'length = 0) then
      done := true;
    else
      read(iline, cur_char);
      case cur_char is
        when ' ' =>
        when ht =>
        when ff =>
        when '#' => comment := true;
                    done := true;
        when others => done := true;
                       strg(1) := cur_char;
      end case;
    end if;
  end loop;
  done := false;
  while (done = false) loop
    if (iline'length = 0) then
      done := true;
    else
      cnt := cnt+1;
      read(iline,strg(cnt));
      if ((strg(cnt) = ' ') or (strg(cnt) = ht) or
         (strg(cnt) = ff)) then
        done := true;
    end if;
    if (cnt > strg'length) then
      done := true;
    end if;
  end if;
end loop;
case strg is
  when "ADD    " => opcode := add;
  when "ADDA   " => opcode := adda;
  when "SUB    " => opcode := sub;
```

```
      when "SUBA    " => opcode := suba;
      when "MUL     " => opcode := mul;
      when "MULA    " => opcode := mula;
      when "DIV     " => opcode := div;
      when "DIVA    " => opcode := diva;
      when "NEGA    " => opcode := nega;
      when "NEGACC  " => opcode := negacc;
      when "STR     " => opcode := str;
      when "RSTR    " => opcode := rstr;
      when "CLEAR   " +> opcode := clear;
      when others =>
          assert false
             report "Illegal opcode in file"
             severity ERROR;
      end case;
  end Read_opcode;
```

Procedure **read_files** reads the external files. This procedure accepts the name of the command file and the answers file, and generates a pin code string to be applied to the UUT pins. The *case* statement is used to decode the opcode and determine if either the *a* argument or the *b* argument or both are required. Then the two argument values are determined accordingly. The opcode and the operands are stored in the **pc_string**.

```
PROCEDURE read_files(
  variable com_file : in TEXT;
  variable ans_file : in TEXT;
  variable pc_string : inout PIN_CODE_STRING) is

  variable opcode : op_type;
  variable iline : LINE;
  variable index : natural;
  variable a, b, c : integer;
  variable a_string : string(1 to 16);
  variable b_string : string(1 to 16);
  variable c_string : string(1 to 32);
  variable a_used : boolean;
  variable b_used : boolean;
  variable ok : boolean;
  variable line_end : natural :=0;
  variable tmp_string : string(1 to input_line_length);
  variable tmp_pc_string : PIN_CODE_STRING;

  BEGIN
    tmp_pc_string := pc_string;
    a_used := true;
```

```
      b_used := true;
      if (endfile(com_file) = false) then
        READLINE(com_file, iline);
        Read_opcode(iline, opcode);
        case opcode is
  when add =>
            pc_string := merge_string(tmp_pc_string, "0000",
                               test_pins'pos(op_code3)+1);
          when adda =>
            pc_string := merge_string(tmp_pc_string, "0001",
                               test_pins'pos(op_code3)+1);
          b_used := false;
          when sub =>
            pc_string := merge_string(tmp_pc_string, "0010",
                              test_pins'pos(op_code3)+1);
          when suba =>
            pc string := merge_string(tmp_pc_string, "0011",
                             test_pins'pos( op_code3)+1);
            b_used := false;
          when div =>
            pc_string := merge_string(tmp_pc_string, "0100",
                               test pins'pos( op_code3)+1);
          when diva =>
            pc_string := merge_string(tmp_pc_string, "0101",
                               test pins'pos( op_code3)+1);
            b_used := false;
          when mul =>
            pc_string := merge_string(tmp_pc_string, "0110",
                                test_pins'pos( op_code3)+1);
          when mula =>
            pc_string := merge_string(tmp_pc_string, "0111",
                                test_pins'pos( op_code3)+1);
            b_used := false;
          when nega =>
            pc_string := merge_string(tmp_pc_string, "1000",
                                test_pins'pos( op_code3)+1);
          b_used := false;
          when negacc =>
            pc_string := merge_string(tmp_pc_string, "1001",
                                test_pins'pos( op_code3)+1);
            a_used := false;
          b_used := false;
          when clear =>
            pc_string := merge_string(tmp_pc_string, "1010",
                                test_pins'pos( op_code3)+1);
              a_used := false;
              b_used := false;
```

```
        when str =>
          pc_string := merge_string(tmp_pc_string, "1100",
                          test_pins'pos( op_code3)+1);
          a_used := false;
          b_used := false;
        when rstr =>
          pc_string := merge_string(tmp_pc_string, "1101",
                          test_pins'pos( op_code3)+1;
          a_used := false;
          b_used := false;
        when others =>
          assert false
            report "Illegal opcode in file."
            severity ERROR;
      end case;
      if (a_used = true) then
        read(iline, a);
        int_to_pin_codes(a,a_string,16);
      pc_string := merge_string(pc_string, a_string,
     test_pins'pos(a15)+1);
   end if;
    if (b_used = true) then
      read(iline, b);
      int_to_pin_codes(b,b_string,16);
      pc_string := merge_string(pc_string, b_string,
     test_pins'pos(b15)+1);
    end if;

    READLINE(ans_file, iline);
    read(iline, c);
    int_to_pin_codes(c,c_string,32);
    pc_string := merge_string(pc_string, c_string,
    test_pins'pos(c31)+1);
  end if;

end read_files;
```

The following is the waveform generation procedure:

```
PROCEDURE waveform(SIGNAL WPL: inout WAVES_PORT_LIST) is

  FILE command_file : TEXT is in "calc_com.txt";
  FILE answer_file : TEXT is in "calc_ans.txt";

  variable pc_string : PIN_CODE_STRING;
```

```
    CONSTANT inputs: pinset:= new_pinset((a15, a14, a13, a12,
      a11, a10, a9, a8, a7, a6, a5, a4, a3, a2, a1, a0, b15,
      b14, b13, b12, b11, b10, b9, b8, b7, b6, b5, b4, b3, b2,
      b1, b0, op_code3, op_code2, op_code1, op_code0));

    CONSTANT outputs: pinset:= new_pinset((c31, c30, c29, c28,
      c27, c26, c25, c24, c23, c22, c21, c20, c19, c18, c17,
      c16, c15, c14, c13, c12, c11, c10, c9, c8, c7, c6, c5,
      c4, c3, c2, c1, c0));

    CONSTANT calc_FSA : Frame_set_array :=
      New_frame_set_array(Non_return(0 ns), inputs) +
      New_frame_set_array(window(1 ns, 10 ns), outputs);

    VARIABLE timing : time_data := new_time_data(calc_fsa);


    BEGIN
      loop
        read_files(command_file, answer_file, pc_string);
        exit when ((endfile(command_file)) or
        (endfile(answer_file)));
        apply(wpl, pc_string, Delay(10 ns), timing);
      end loop;
  END waveform;
END WGP_Calculator;
```

The waveform generation procedure generates stimulus and response data in frames of 10 ns. The inputs to the calculator are generated at the beginning of each slice and response is expected in a window from 1 ns from the beginning of the slice until the end of the slice. The stimulus and response data are read respectively from the command and answer files using the **read_files** procedure. All of the above procedures reside in the body of a package whose declaration is shown below:

```
PACKAGE WGP_Calculator is
  PROCEDURE int_to_pin_codes(variable num : in integer;
                             variable pc_string : out string;
                             constant len : in integer);

  PROCEDURE Read_opcode(variable  iline : inout line;
                        variable opcode : out op_type);

  function strcmp( str1 : in string ; str2 : in string ) return
    boolean;

  PROCEDURE read_files(
```

```
    variable com_file : in TEXT;
    variable ans_file : in TEXT;
    variable pc_string : inout PIN_CODE_STRING);
  PROCEDURE waveform(SIGNAL WPL: inout WAVES_PORT_LIST);
END WGP_Calculator;
```

The VHDL test bench for the calculator examples is shown below:

```
LIBRARY ieee;
USE ieee.std_logic_1164.ALL;

LIBRARY waves_std;
USE waves_std.WAVES_SYSTEM.all;

LIBRARY waves_1164;
USE waves_1164.WAVES_1164_utilities.all;
USE waves_1164.WAVES_1164_logic_value.all;

USE WORK.UUT_test_pins.all;
USE work.waves_objects.all;

USE work.WGP_calculator.all;
USE work.constants.all;
USE STD.textio.all;
USE ieee.std_logic_textio.all;

ENTITY test_bench IS

END test_bench;



ARCHITECTURE calculator_test OF test_bench IS

function wpl_2_op_type( Port_list : in system_waves_port_list)
                        RETURN op_type is

  variable tmp_vect : std_logic_vector(Port_list'range);
  variable result : op_type;

BEGIN

  tmp_vect := stim_1164(Port_list);
  for I in tmp_vect'range loop
    if tmp_vect(i) = '-' then
      tmp_vect(i) := '0';
    end if;
```

```
  end loop;
  if (tmp_vect = "0000") then return add;
  end if;
  if (tmp_vect = "0001") then return adda;
  end if;
  if (tmp_vect = "0010") then return sub;
  end if;
  if (tmp_vect = "0011") then return suba;
  end if;
  if (tmp_vect = "0100") then return div;
  end if;
  if (tmp_vect = "0101") then return diva;
  end if;
  if (tmp_vect = "0110") then return mul;
  end if;
  if (tmp_vect = "0111") then return mula;
  end if;
  if (tmp_vect = "1000") then return nega;
  end if;
  if (tmp_vect = "1001") then return negacc;
  end if;
  if (tmp_vect = "1010") then return clear;
  end if;
  if (tmp_vect = "1100") then return str;
  end if;
  if (tmp_vect = "1101") then return rstr;
  end if;
  assert false
    report "Illegal opcode."
    severity ERROR;
  return add;
end wpl_2_op_type;


  PROCEDURE int_2_std_logic(num : in integer;
                           variable out_vector  : out
                           std_logic_vector) is
    variable tmp : integer;
    variable orig : integer;
    variable cnt : integer;
    variable tmp_vector : std_logic_vector(out_vector'range);
    variable len : natural;

  BEGIN
    len := out_vector'length;
    tmp := num;
```

```
    orig := num;
    for I in 0 to len-1 loop
      if (tmp mod 2 = 1) then tmp_vector(i) := '1';
      else tmp_vector(i) := '0';
      end if;
      tmp := tmp / 2;
    end loop;
    if (orig<0) then
      for i in out_vector'range loop
        if (tmp_vector(i) = '0') then
          tmp_vector(i) := '1';
        else tmp_vector(i) := '0';
        end if;
      end loop;
      cnt := 0;
      loop
        if (tmp_vector(cnt) = '0') then
          tmp_vector(cnt) := '1';
        else tmp_vector(cnt) := '0';
        end if;
        exit when ((tmp_vector(cnt) = '1') or (cnt =
          out_vector'length));
        cnt := cnt + 1;
      end loop;
    end if;
    out_vector := tmp_vector;
  End int_2_std_logic;


function wpl_2_int( Port_list : in system_waves_port_list)
                    RETURN integer is

  variable tmp_vect : std_logic_vector((Port_list'right-
    Port_list'left)
                                         downto 0);
  variable ret_val  : integer := 0;
  variable bit_chk  : std_logic   := '1';
  variable negate   : integer := 1;
  variable incr     : integer := 0;

BEGIN

  tmp_vect := stim_1164(Port_list);
  if(tmp_vect(tmp_vect'HIGH) = '1') then
    bit_chk := '0'
    negate := -1;
```

```
    incr := 1;
  end if;

  for I in tmp_vect'RANGE loop
    if(tmp_vect(i) = bit_chk) then
      ret_val := 2**i + ret_val;
    end if;
  end loop;

  return negate*(ret_val+incr);
END wpl_2_int;

  --*************************************************************
  --**************CONFIGURATION SPECIFICATION****************
  --*************************************************************

   COMPONENT calculator
    PORT ( A                 : IN   integer;
           B                 : IN   integer;
           opcode            : IN   op_type;
           C                 : OUT  integer);
    END COMPONENT;

 -- Modify entity use statement
 -- User Must Modify modify and declare correct
 --   .. Architecture, Library, Component ..
 -- Modify entity use statement
FOR ALL:calculator USE ENTITY work.calculator(behav);

  --*************************************************************
  -- stimulus signals for the waveforms mapped into UUT INPUTS
  --*************************************************************

    SIGNAL WAV_STIM_A                  :integer := 0;
    SIGNAL WAV_STIM_B                  :integer := 0;
    SIGNAL WAV_STIM_opcode             :op_type := clear;

  --*************************************************************
  -- Expected signals used in monitoring the UUT OUTPUTS
  --*************************************************************

    SIGNAL FAIL_SIGNAL          : std_logic;
    SIGNAL WAV_EXPECT_C         : std_ulogic_vector(31 downto 0);

  --*************************************************************
  -- UUT Output signals used in Monitoring ACTUAL Values
  --*************************************************************
```

```
    SIGNAL ACTUAL_C           : integer;

  --**********************************************************
  -- Bi_directional signals used for stimulus signals mapped
  -- into UUT INPUTS and also monitoring the UUT OUTPUTS
  --**********************************************************


--
--
-- No Bidirectional Pins On UUT



  --**********************************************************
  -- WAVES signals OUTPUTing each slice of the waves port list
  --**********************************************************

        SIGNAL wpl   : WAVES_port_list;

BEGIN
  --
  --**********************************************************
  -- process that generates the WAVES waveform
  --**********************************************************

       WAVES: waveform(wpl);

  --**********************************************************
  -- processes that convert the WPL values to 1164 Logic Values
  --**********************************************************

  WAV_STIM_A             <= wpl_2_int(wpl.wpl( 1 to 16 ));
  WAV_STIM_B             <= wpl_2_int(wpl.wpl( 17 to 32 ));
  WAV_STIM_opcode        <= wpl_2_op_type(wpl.wpl( 33 to 36 ));
  WAV_EXPECT_C           <= EXPECT_1164(wpl.wpl( 37 to 68 ));


  --****************************************
  -- UUT Port Map - Name Symantics Denote Usage
  --****************************************

  u1: calculator
  PORT MAP(
    A                       => WAV_STIM_A,
    B                       => WAV_STIM_B,
```

```
    opcode                    => WAV_STIM_opcode,
    C                         => ACTUAL_C);

  --************************************************************
  -- Monitor Processes to Verify The UUT Operational Response
  --************************************************************

Monitor_C:
  PROCESS(ACTUAL_C, WAV_expect_C)
    variable vector_c : std_logic_vector(31 downto 0);

  BEGIN
       int_2_std_logic(actual_c, vector_c);
       assert(Compatible (actual => vector_c,
                         expected => WAV_expect_c))
       report "Error on c output" severity WARNING;

  IF ( Compatible ( vector_c,   WAV_expect_c) ) THEN
    FAIL_SIGNAL <='L'; ELSE FAIL_SIGNAL <='1';
  END IF;

  END PROCESS


END calculator_test;
```

10.5 Summary

In ***summary of our chapter on WAVES Level 2***, the WAVES specification language contains constructs discussed earlier in the book with additional language constructs, i.e. WAVES Level 2. While Level 1 is sufficient for most simple test environments, it is insufficient when more general test methods are required. WAVES Level 2 extends WAVES Level 1 through the addition of a number of high-level programming constructs from VHDL. These constructs include:

- Extended type declarations
- Variable assignments within subprograms
- More flexible file I/O

In this chapter we provided a review of each of these enhancements, and portions of two WAVES datasets were shown, illustrating how each of these enhancements are put to use. The calculator example from this chapter is included in the **calc** directory on the companion CD-ROM.

CHAPTER 11. INTERACTIVE WAVEFORMS: HANDSHAKING AND MATCHING

Actively interacting with the unit under test

In this chapter of our advanced topics portion of this book, we move beyond the simple stimulus-response paradigm we have presented so far, and outline how WAVES may be used in an interactive fashion to test a UUT. In the previous chapters, we have discussed the use of WAVES in generating waveform events independently of the UUT behavior. In such typical generate-apply applications, the waveform generation procedure is completely asynchronous with respect to the UUT. It simply generates the waveforms and the test bench takes care of the proper application of the stimuli data to the UUT pins and the comparison of the actual UUT responses with the expected responses, if any, recorded in the WAVES-generated waveforms.

In contrast, there are occasions where we may wish to generate *interactive* waveforms -- waveforms whose events depend on the UUT responses. Such waveforms are called *interactive waveforms* since they interact with the UUT. The need for interactive waveforms arises in the following typical applications:

- For UUTs whose initial states are unknown, it may be necessary to cycle the device through an unknown number of clock cycles until the device settles into a known state indicated by the outputs. In this case, the waveform has to contain an arbitrary number of clock pulses, until the UUT reaches the desired state.

- For self-clocking devices, it is sometimes necessary for the stimulus waveform to 'wait' for a specific response from the UUT before the next waveform event can be generated and applied. In this case, the delay between successive events in the stimulus waveform is not fixed, but rather varies, and depends on the UUT response to the previous inputs.

- For UUTs with asynchronous interfaces, it is necessary for the device to respond to a previous request before the next request can be issued. Again, the delay between successive requests is a variable and is dependent upon the UUT response time which, in general, is unknown *a priori*.

WAVES provides two primary ways of generating interactive waveforms. The first technique is based on *handshake delays*. The second technique is based on *match* flags and is used in conjunction with the match modes commonly found in Automated Test Equipment (ATE). We introduce and discuss both of these techniques, and provide some examples of their use, in this chapter.

11.1 Handshake Delays

The first method of generating an interactive waveform is by using handshake delays in the APPLY procedure calls, within the waveform generation procedures. Recall that the APPLY procedures provided by the WAVES standard packages have two basic purposes: (1) to generate waveform events, and (2) to advance time. The time to be advanced following the application of the event data to the waveform is typically specified through a **Delay_Time** type argument. Recall that the **Delay_Time** has to be supplied either as an explicit argument or through a **Wave_Timing** construct.

11.1.1 APPLY Procedures

We will consider the following variant of the APPLY procedure where the **Delay_Time** is passed as an explicit argument:

```
procedure APPLY(
 signal   CONNECT   :inout Waves_Port_List;
 signal   MATCH     :in   Waves_Match_List;
 constant CODES     :in    Pin_Code_String;
 constant DELAY     :in    Delay_Time;
 variable FRAMES    :in    Time_Data;
 constant ACTIVE_PINS :in    Pinset := All_Pins);
```

In comparison with the APPLY procedures we discussed in the previous chapters, we note two arguments here. The first, the **Delay_Time** type argument, can be either a timed delay or a handshake delay. We have discussed fixed-time delays in the previous chapters and will discuss handshake delay in this chapter. The second argument that should be noted is *Match* which is of the type **Waves_Match_List**. This argument is required when a handshake delay is used in the **Delay_Time** construct. We will discuss this argument further later in this section. Let's start with the **Delay_Time** argument.

The **Delay_Time** construct is used to pass a delay object as an argument to the APPLY procedure. The **Delay_Time** value governs the time to elapse prior to the time of the next APPLY operation. As we have seen in the previous chapters, this value typically comes from the time entry in the external file that contains the pin code data and the time data for the waveforms. However, if we wish to apply a handshake delay, that is, we wish to suspend the next APPLY operation until the UUT responds by generating a specified value on a specified output pin, we can specify a handshake delay in the **Delay_Time** construct.

Delay_Time type objects are generated using *Delay* function calls. Some variants of the *Delay* function facilitate the specification of handshake information. Consider the following variant provided by the WAVES standard packages:

```
function Delay(
 NOMINAL        : TIME;
 BASE_LOGIC      : Logic_Value;
 BASE_PIN       : Test_Pins)
 return Delay_Time;
```

Here, **BASE_PIN** specifies the output pin of the UUT on which the handshaking event is expected to take place. **BASE_LOGIC** specifies the logic value which is expected to appear on the base pin. More specifically, the event expected is that of a *transition* on the base pin to the **BASE_LOGIC** value from some other logic value. It is that transition which satisfies the required handshake. The next APPLY operation is suspended until the UUT produces this handshake event. *NOMINAL* specifies the time after the handshake event until which the application of the next slice is delayed. This is usually set to 0 ns, in which case the waveform time is advanced to the time of the handshake event and the next slice is immediately applied without any further delay.

An example use of the above *Delay* function call is as follows:

```
constant HandShake: Delay_Time := Delay(5 ns, SENSE_1, Ack);
```

In this example, *HandShake* specifies a handshake event on the *Ack* pin. *Ack* is expected to transition to the logic value **Sense_1** from some other value. Note that if *Ack* is already at **Sense_1**, then it is necessary for it to change to some other value *and then change back* to **Sense_1** in order for handshaking to be successful. In other words, an edge on *the Ack* pin whose final value is **Sense_1** is expected. In this example, *5 ns* specifies that the next slice starts 5 ns after the handshaking edge occurs on the *Ack* pin.

11.1.2 Waveform Generation Procedure

When handshake delays are used, the waveform generation procedure requires a second parameter, namely the **Waves_Match_List**, which should immediately follow the first **Waves_Port_List** parameter:

```
procedure waveform(signal wpl: inout WAVES_port_list;
          signal wml: in WAVES_match_list);
```

The **Waves_Match_List** has a field called **D_Flag** which is a boolean flag. This flag is set by the environment (ATE or VHDL test bench) to indicate the occurrence of the handshake event in the environment. This flag must transition from false to true in order for the handshaking to be recognized by the WAVES dataset. The environment must cause a false to true transition on this flag when successful handshaking takes place. We will use an example test bench to illustrate how this is done.

11.1.3 Example

Consider a simple adder that adds two input bits, A and B, and produces two output bits, Sum and Carry. The delay through the adder is unknown and varies depending upon the inputs. To allow interaction with the environment, our adder also has two additional interface signals: Start and Finish. When an addition operation is required, the environment should set A and B values and place a rising edge on the Start input pin. The adder first pulls the Finish signal to low and starts the addition. When the addition is completed, the adder generates the Sum and Carry outputs and places a rising edge on the Finish output pin, to let the environment know that the result is ready. We present a sketch of the waveform generation procedure to test the Adder. We wish to apply the four possible combinations of the A and B values to the adder and examine the outputs.

Waveform Generation Procedure - Our overall plan is to set the inputs to the desired values and pulse the Start signal. We will wait until the Finish goes high. All this happens in the first slice, which should have a handshake delay. When the Finish goes high, we will start the second slice with a fixed duration of 10ns. In this slice, we will create a 5 ns window during which we will check that Finish is indeed high and that the outputs have the expected values. We will repeat this process for all the desired combinations of input values. The following is the sketch of our waveform generation procedure:

```
PROCEDURE waveform(signal Wpl: inout, WAVES_PORT_LIST,
          signal Wml: in, WAVES_MATCH_LIST) is
```

```
CONSTANT inputs: pinset := new_pinset((A, B));
CONSTANT outputs: pinset := new_pinset((Carry, Sum));

CONSTANT FSA1 : Frame_set_array :=
 New_frame_set_array(Pulse_High(1ns, 2ns), Start) +
 New_frame_set_array(Non_return(0ns), inputs) +
 New_frame_set_array(Window(0ns, 0ns), Finish) +
 New_frame_set_array(Window(0ns, 0ns), outputs);

CONSTANT FSA2 : Frame_set_array :=
 New_frame_set_array(Non_return(0ns, 5ns), Start) +
 New_frame_set_array(Non_return(0ns), inputs) +
 New_frame_set_array(Window(0ns, 5ns), Finish) +
 New_frame_set_array(Window(0ns, 5ns), outputs);

VARIABLE Timing1: time_data:= new_time_data(FSA1);
VARIABLE Timing2: time_data:= new_time_data(FSA2);

constant Finished: Delay_Time := Delay(0 ns, SENSE_1, Finish);

-- Order of pins: Start A B Finish Carry Sum

Apply(Wpl, Wml, "100---", Finished, Timing1);
Apply(Wpl, Wml, "000100", Delay(10 ns), Timing2);

Apply(Wpl, Wml, "101---", Finished, Timing1);
Apply(Wpl, Wml, "001101", Delay(10 ns), Timing2);

Apply(Wpl, Wml, "110---", Finished, Timing1);
Apply(Wpl, Wml, "010101", Delay(10 ns), Timing2);

Apply(Wpl, Wml, "111---", Finished, Timing1);
Apply(Wpl, Wml, "011110", Delay(10 ns), Timing2);

end waveform;
```

First, we declare two pinsets: *inputs* containing input pins A and B and *outputs* containing output pins Carry and Sum. We then declare two frameset arrays *FSA1* and *FSA2.FSA1* schedules a **Pulse_High** event sequence on the Start pin and **Non_Return** event sequences on the inputs. The windows placed on Finish and outputs are not relevant. *FSA2* has **Non_Return** event sequences on Start and inputs and 5 ns windows on Finish and the outputs. The idea is to strobe the Finish and output signals during this window. We then declare two **Time_Data** objects Timing1 and Timing2 containing the frameset arrays FSA1 and FSA2 respectively.

We then declare a **Delay_Time** object, Finished, which defines our handshake event. Finished declares the handshake event to be that of Finish signal going to **SENSE_1** state from a non **SENSE_1** state. 0 ns is the time between the handshake event and the beginning of the next slice.

We then have a sequence of slice applications using Apply procedure calls. These calls are shown in pairs. The first Apply in each pair applies the desired input values and a high pulse on the Start signal, with a handshake delay on the Finish signal. When Finish goes high, the second Apply in the pair applies the next slice which schedules strobes on Finish and the two output pins, holding Start at the low logic level. This process is repeated for all the four combinations of input values A and B.

VHDL Test Bench - When the WAVES dataset is used in conjunction with WAVES-compatible Automated Test Equipment (ATE), the ATE should provide proper interface between the WAVES system and the ATE operating system to set various internal signals to handle handshaking properly. However, in the case of VHDL test benches, it is necessary for the user to write a special VHDL process in the test bench to manage handshaking. Following is one way of writing a test bench process for this purpose.

```
signal wpl: Waves_Port_List;
signal wml: Waves_Match_List;

WAVES: waveform(Wpl, Wml);  -- concurrent procedure call on the
              -- waveform generator.

HandShake_Process: PROCESS
 BEGIN
 wml.d_flag <= true;
 wait until wpl.delay_flag = true;

 case wpl.delay_pin is
  when 4 =>
  if UUT_Finish = To_1164(Logic_Value'val(wpl.delay_logic-1))
    then wml.d_flag <= true;
    else wml.d_flag <=false;
  end if;

  when 5 =>
   -- similar code for UUT_Carry
  when 6 =>
   -- similar code for UUT_Sum
  when others =>
   assert false
```

```
    report "Attempting handshake on a non-output pin of UUT."
    severity ERROR
 end case;

 loop
  case wpl.delay_pin is
   when 4 =>
   wait on UUT_Finish;
   if UUT_Finish = To_1164(Logic_Value'val(wpl.delay_logic-1))
    then wml.d_flag <= true;
    else wml.d_flag <= false;
   end if;

  when 5 =>
   -- similar code for UUT_Carry
  when 6 =>
   -- similar code for UUT_Sum
  when others =>
   assert false
    report "Attempting handshake on a non-output pin of UUT."
    severity ERROR
 end case;

 wait for 0ns;
 wait for 0ns;

 exit when wpl.delay_flag = false;
 end loop;
END PROCESS Handshake_Process;
```

The test bench contains a concurrent procedure call on the waveform generation procedure with the WAVES port list and the WAVES match list as the arguments. These two parameters facilitate interaction between the WAVES dataset and the test bench. The **D_Flag** entry in the **Waves_Match_List** structure defined in the WAVES packages is a boolean flag. This is set by the environment and read by the WAVES dataset. The handshake process in our test bench begins by setting this flag to true. It then suspends until the **Delay_Flag** in the WAVES port list becomes true. The **Delay_Flag** in the port list is used by the WAVES dataset to indicate to the environment that a handshake delay has been scheduled and the waveform generator is suspended, waiting for a false to true transition on **wml.d_flag**.

When the **wpl.delay_flag** becomes true, the handshake process in the testbench recognizes that the WAVES dataset is waiting for a handshake. The handshake process (or the environment in general) has to now determine the identification of the handshake pin and its expected handshake value. These two pieces of information are recorded in **wpl.delay_pin** and **wpl.delay_logic** respectively. **Wpl.delay_pin** is an integer value which indexes into the **Test_Pins** enumeration and denotes the UUT pin on which handshaking is desired. **Wpl.delay_logic** likewise is an integer that indexes into the **Logic_Value** enumeration (with an offset of 1 due to the way the WAVES standard packages are written). The logic value at that position in the enumeration denotes the value expected on the designated pin for handshaking to be successful.

The case statement has a branch for each possible value of the **Delay_Pin**. Handshaking is permitted on any output pin of the UUT. For each output pin, the actual current value of the UUT pin (this value being visible in the testbench) is compared with the expected handshake value and the **wpl.d_flag** is set to true or false accordingly. Recall that the handshaking is successful only when this signal transitions from false to true. **To_1164** is a function (not shown) defined in the testbench to convert **Logic_Value** enumeration into IEEE standard 1164 logic value enumeration, to facilitate the comparison of the expected and actual values.

After thus determining the initial value of the **wml.d_flag**, the handshake process enters a loop which is exited only when the handshaking is completed successfully, that is, when the **wml.d_flag** transitions from false to true and that transition is recognized by the waveform generation procedure. The case statement in the loop again branches based on the handshake pin. The process waits for an event on the handshake pin. Following the event, it compares the expected and actual values on the pin and accordingly sets the **wml.d_flag**. The 0 ns wait statements at the end of the case statement make the new value of **wml.d_flag** visible to the WAVES dataset in the next simulation cycle.

If the **wml.d_flag** transitioned from false to true, then the WAVES dataset recognizes this to be a successful handshake and responds by setting **wpl.delay_flag** to false, thereby removing the pending handshake delay, and gets ready for the application of the next slice. This, in turn, causes the exit condition of the loop statement to be successful and the handshake process goes back to the top, sets the **wml.d_flag** to true (it should already be true since handshaking was just successful) and waits for the scheduling of the next handshake event by the WAVES dataset.

If the **wml.d_flag** did not transition from false to true, then the loop body repeats until this flag first becomes false and then becomes true. Only such a positive transition is recognized as successful handshake by the WAVES dataset. Until that happens, the WAVES dataset will not set the **wpl.delay_flag** to false and hence, the loop body repeats.

11.2 Matching

Another way of generating interactive waveforms is to use the Match operations to determine whether the UUT produced an expected output and then to proceed based on the outcome of the test. This is compatible with the match mode functionality supported by Automated Test Equipment (ATE). To support matching, WAVES assumes that the environment (either the VHDL testbench or the ATE) supports *match flags*. We will begin our discussion of matching with the concept of match flags.

To support match mode datasets, the environment must provide two registers for each pin in **Test_Pins**. The first register is called the *control* register or *control flag*. The control flag stores a one-bit flag that assumes the values Sample and Hold. The second register is called the *match result* register or *match flag*. The *match flag* which assumes the values True and False, indicates whether matching on that pin was successful or not. Each pin has an associated comparator in the environment (ATE or testbench) that, when turned on through the use of the control flags, compares the expected and actual values on the pin and sets the match flags accordingly. These match flags can then be sampled by the waveform generator through a pre-defined *Match* function provided for that purpose.

Here is how matching works:

1. Initially, all match flags should be set to False and all control flags should be set to Hold.

2. When matching is desired on certain pins, the waveform generator turns on the corresponding comparators by setting their control flags to Sample. This is done by a Match procedure call which we explain later in this section.

   The environment responds by setting the match flags to True and turning on comparison on those pins. The comparators do nothing unless the event value has a direction of *response* and a relevance of *required*. In other words, matching is relevant only for required response type events which denote required outputs of the UUT. Then the comparators should determine whether the state and strength of the actual output of the UUT matches the expected values on the waveform. If there is a mismatch, then the comparator sets the match flag of the corresponding pin to False. Otherwise, the match flag will continue to be True. This matching is done continuously until stopped by turning the comparators off as we explain in the next step.

3. When the waveform generator desires to stop the matching process on certain pins, it requests the environment to set the control flags of those pins to Hold. This is done by a Match *procedure* call as we will explain shortly. The environment responds by setting the control flags of those pins to Hold. Then, the comparison stops until the next time the control flags are set to Sample, and the current match flags hold their values for possible inspection by the waveform generator.

4. The waveform generator can inspect the match flags at any time (regardless of the control flag values, although inspection is usually done when the control flags are set to Hold) by using the Match *functions* to be introduced shortly. The Match functions take one or more pins and return the Boolean True if all the corresponding match flags are set to True. Otherwise, the Match functions return False.

Interaction between the match mode comparators in the environment and the WAVES dataset is facilitated through the Match procedures and functions defined by the WAVES standard. The Match procedures are used to initiate and terminate matching by setting the control flags, and the Match functions are used to test the match flags. We describe these next.

11.2.1 Match Procedures

There are two Match procedures, one for controlling matching on a single pin and the other for handling multiple pins. Both versions are shown below:

```
procedure Match(
 signal   CONNECT    :out Waves_Port_List;
 constant CONTROL    :in  Match_Control_Type;
 constant ACTIVE_PIN :in  Test_Pins);

procedure Match(
 signal   CONNECT    :out Waves_Port_List;
 constant CONTROL    :in  Match_Control_Type;
 constant ACTIVE_PINS:in  Pinset := All_Pins);
```

The **Waves_Port_List** argument is already familiar to us. **Match_Control_Type** is defined by the WAVES standard packages as the following enumerated type:

```
type Match_Control_Type is (HOLD, SAMPLE);
```

Accordingly, the CONTROL argument to the Match procedures specifies whether the control flags should be set to Sample or to Hold, thereby initiating or terminating the matching process. The last argument identifies one or more pins involved in the matching operation.

11.2.2 Match Functions

There are two Match functions, one for testing the match flag of a single pin and the other for testing the match flags of a set of pins, provided by the WAVES packages. These are as follows:

```
function Match(
 constant CONNECT   : Waves_Match_List;
 constant ACTIVE_PIN: Test_Pins)
 return BOOLEAN;

function Match(
 constant CONNECT   : Waves_Match_List;
 constant ACTIVE_PINS: Pinset := All_Pins)
 return BOOLEAN;
```

The first argument to the Match functions identifies the **Waves_Match_List**. The second argument identifies a single pin or a set of pins involved in matching. In the first case, the Match function returns the current value of the match flag for that pin. In the second case, the Match function returns true if current values of the match flags of all the **Active_Pins** are true. Otherwise, it returns false. In other words, matching fails if any one of those match flags is false. The role of the **Waves_Match_List** will become clear later in this section.

11.2.3 APPLY Procedures

APPLY procedures used in match mode waveforms must have a **Waves_Match_List** parameter. The following is one variant of the APPLY procedure for match mode waveforms:

```
procedure APPLY(
 signal   CONNECT    :inout Waves_Port_List;
 signal   MATCH     :in   Waves_Match_List;
 constant CODES      :in   Pin_Code_String;
 constant DELAY      :in   Delay_Time;
 variable FRAMES     :in   Time_Data;
 constant ACTIVE_PINS :in   Pinset := All_Pins);
```

11.2.4 Waveform Generation Procedure

In the previous chapters, we discussed one parameter of the waveform generation procedure, namely, the **Waves_Port_List**. Waveform generation procedures for match mode waveforms require a second parameter, namely, the **Waves_Match_List** which should immediately follow the first **Waves_Port_List** parameter. Hence, for match mode waveforms, the waveform generation procedure declaration might appear as follows:

```
procedure waveform(signal wpl: inout WAVES_port_list;
          signal wml: in WAVES_match_list);
```

The **Waves_Match_List** has a field called **M_Flags** which is an array of boolean flags, one for each pin in the **Test_Pins** enumeration. These denote the values of the match flags and should be maintained by the environment (the ATE in case of hardware testing or the VHDL testbench in case of simulation). The role of the **Waves_Match_List** will become clear through the following example.

11.2.5 Example

Let us consider our adder example again. This time, we assume that the adder has inputs A and B and outputs Carry and Sum. It has no Start and Finish signals. This time we would like to supply the desired input values and sustain these input values until the outputs match the expected values.

Waveform Generation Procedure - The following shows the sketch of the waveform generation procedure.

```
PROCEDURE waveform(signal wpl: inout, WAVES_PORT_LIST,
          signal wml: in, WAVES_MATCH_LIST) is

CONSTANT inputs: pinset := new_pinset((A, B));
CONSTANT outputs: pinset := new_pinset((Carry, Sum));

CONSTANT FSA : Frame_set_array :=
 New_frame_set_array(Non_return(0ns), inputs) +
 New_frame_set_array(Window(0ns, 1ns), outputs);

VARIABLE Timing: time_data:= new_time_data(FSA);

-- Order of pins: A B Carry Sum

loop
 MATCH(wpl, SAMPLE, outputs);
```

```
 APPLY(wpl, wml, "0000", Delay(1 ns), Timing);
 MATCH(wpl, HOLD, outputs);
 exit when MATCH(wml, outputs);
end loop;

loop
 MATCH(wpl, SAMPLE, outputs);
 APPLY(wpl, wml, "0101", Delay(1 ns), Timing);
 MATCH(wpl, HOLD, outputs);
 exit when MATCH(wml, outputs);
end loop;

loop
 MATCH(wpl, SAMPLE, outputs);
 APPLY(wpl, wml, "1001", Delay(1 ns), Timing);
 MATCH(wpl, HOLD, outputs);
 exit when MATCH(wml, outputs);
end loop;

loop
 MATCH(wpl, SAMPLE, outputs);
 APPLY(wpl, wml, "1110", Delay(1 ns), Timing);
 MATCH(wpl, HOLD, outputs);
 exit when MATCH(wml, outputs);
end loop;
end waveform;
```

We begin by declaring the inputs and outputs pinsets. We then define a **Time_Data** object containing **Non_Return** event sequences for the inputs and window sequences for the outputs spanning the entire duration of the slice. The waveform generation procedure contains four loop statements, each testing for one of the input combinations. We will consider the first loop statement. The first MATCH statement in the loop turns matching on by making a MATCH procedure call with SAMPLE as an argument. This turns on the comparators in the environment for the two output pins. The APPLY statement immediately applies a slice, setting the input values to 00 and the output values also to 00 for the entire 1 ns duration of the slice. In the environment, matching is taking place continuously for the duration of the slice. Note that matching is effective only for the output pins for which the direction is *response* and the relevance is *required.* At the end of the slice, the waveform generator makes a second call to the MATCH procedure with the parameter HOLD to hold the values of the match flags and turn the comparators in the environment off. The Match function call checks whether the match flags for all the output pins are held at true. If so, the actual outputs matched the expected outputs and testing can continue to the next set of input values. If not, the loop body iterates to attempt matching over another 1 ns slice for the same stimuli values.

VHDL Testbench - As usual, if the WAVES system is installed on a piece of ATE, then the match flags will be maintained automatically by the ATE and the WAVES system interface. However, if the environment for WAVES execution is a VHDL simulator, then the testbench needs to manage the match flags explicitly. The following is a process that can be included in the testbench, illustrating one way to manage the match flags; this process manages the match flags of the Carry pin.

```
Carry_match_maintainer: process
 begin
  wml.m_flags(3) <= false;
  loop
   wait until wpl.wpl(3).m_control = SAMPLE_START;
   wml.m_flags(3) <= true;
   wait until wpl.wpl(3).m_control = SAMPLE;
   wml.m_flags(3) <= wml.m_flags(3) and (Compatible(actual =>
Carry,
            expected => expect_1164(wpl.wpl(3))));
   loop
     wait on Carry, wpl;
     wml.m_flags(3) <= wml.m_flags(3) and (Compatible(actual =>
Carry,
              expected =>expect_1164(wpl.wpl(3))));
     exit when wpl.wpl(3).m_control = HOLD;
   end loop;
  end loop;
 end process Carry_match_maintainer;
```

We need to included as many processes of the above type in the testbench as the number of pins for which match flags need to be maintained. The above process maintains the match flag for the Carry pin which is pin number 3 in the **Test_Pins** enumeration.

Recall that the **Waves_Match_List** structure has a set of match flags (**M_Flags**), one for each test pin. Initially, the environment, our test bench process in this case, sets this flag to false. Then this process enters a loop that repeats forever. Inside this loop, the process waits for the **M_Control** (match control) flag of the Carry pin to become **SAMPLE_START**. Match control flags are available in the **Waves_Port_List** structure. If the pin's match control flag is set to **SAMPLE_START** then its match flag is turned on. Then the match process waits for the control flag to become SAMPLE. Then the match flag is updated by adding its previous value with the result of comparing the current value of the Carry pin with its expected value. The comparison continues until the control flag becomes HOLD.

We need to understand this in conjunction with what happens in the WAVES dataset. When the MATCH procedure call with the SAMPLE parameter is called in the waveform generation procedure, the WAVES dataset sets the match control (**M_Control**) of the match pins to **SAMPLE_START**. The environment (our testbench process) responds by setting the match flags to true. The WAVES dataset then changes the match controls to the SAMPLE state. The environment responds by beginning the comparison process and continues the comparison whenever the expected or actual UUT output values change and as long as the match controls are in the SAMPLE state. When the waveform generator procedure calls the MATCH procedure with the HOLD parameter, the environment responds by not performing further comparisons and thereby holding the match flag values.

Matching is an alternative to our earlier way of writing test benches using explicit comparison. Match mode waveforms are portable to ATE, whereas the non-match mode waveforms with explicit comparison in testbench are not portable with respect to the comparison since testbenches are not portable to ATE.

In ***Summary of Chapter 11***, we introduced two methods of writing interactive waveform generators. When coarse handshaking is needed, match mode waveforms can be used to set windows of matching during which comparison between expected and actual waveforms is done automatically. However, when more precise handshaking is desired, we must use handshake delays. Precise handshaking through handshake delays is limited to a single pin whereas matching using match mode waveforms can be done on a number pins at the same time. Also, we note that it is possible to combine matching and handshaking features in interesting ways in a WAVES dataset to generate interactive waveforms.

CHAPTER 12. USING WAVES FOR BOUNDARY-SCAN ARCHITECTURES

Serial test vector applications

In the previous chapters, we have concentrated on using WAVES for generating parallel tests, i.e. each test vector is applied to the device pins in parallel. In this chapter, we will examine how WAVES can be used to generate serial test sequences, i.e. test vectors that should be applied serially to certain pins of the device under test. Such capability is important while testing devices that are IEEE 1149.1 boundary-scan standard compliant.

12.1 Boundary-Scan Architecture

The IEEE 1149.1 standard defines a boundary-scan architecture and a test access port configuration to exercise the scan circuitry on the device [1,2]. IEEE 1149.1 compliant devices are testable through the serial interface. The serial interface consists of four mandatory pins, Test Clock (TCK), Test Mode Select (TMS), Test Data Input (TDI), and Test Data Output (TDO), and one optional pin, Test Reset (TRST). The precise usage of these pins is not important for understanding the material in this chapter. Readers interested in the 1149.1 architecture should consult the references [1,2].

For the purposes of this chapter, it is sufficient to know that a boundary-scan capable device has a number of pins on which serial data is expected (serial pins) and a number of pins on which parallel data may be applied (parallel pins). A typical test involves the application of a small number (usually one or two) of parallel vectors, followed by a relatively large number of serial vectors, through the serial pins (usually the four pins mentioned before: TCK, TMS, TDI and TDO). This process may be repeated several times in order to apply different tests.

12.2 External File Conventions

For the purposes of this chapter we will assume that parallel and scan vectors can be intermixed. We will also assume that the pin codes are stored in an external file in the WAVES external file format using the following conventions:

- Scan vectors are written with a file slice integer value of one. Scan vectors consist of up to 5 pin codes and map to the scan test pins in the following order: TCK, TMS, TDI, TDO and TRST (if present).
- Vectors are written in the usual manner but must have a file slice integer greater than one.

The following is an example of an external file:

```
%PIO BEGIN
01010101010101010101010101010 : 2;
%TRST ON
=5 0 : 1;
%From Stable state: Undefined to Shift state: DRSHIFT
10-- : 1;
10-- : 1;
10-- : 1;
11-- : 1;
11-- : 1;
10-- : 1;
10-- : 1;
%DR IR Command.
100- : 1;
101- : 1;
100- : 1;
101- : 1;
100- : 1;
101- : 1;
100- : 1;
101- : 1;
%DR IR Command.
101- : 1;
100- : 1;
100- : 1;
100- : 1;
100- : 1;
100- : 1;
100- : 1;
100- : 1;
100- : 1;
100- : 1;
100- : 1;
100- : 1;
100- : 1;
100- : 1;
100- : 1;
```

```
100- : 1;
100- : 1;
100- : 1;
111- : 1;
%From Exit State: DREXIT1 to Stable State: IDLE
11-- : 1;
10-- : 1;
%PIO BEGIN
Z01-00010101010101010101010101010 : 2;
```

In this example, the first and last lines in the file denote parallel vectors. The remaining entries denote serial inputs. Through this file, the user intends to apply a parallel vector, turn the TRST input ON, set the scan circuitry into the so-called DRSHIFT mode, issue a sequence of serial inputs to the scan circuitry, and finally apply a parallel vector.

12.3 Waveform Generation Procedure

Our overall approach in writing the waveform generator procedure (WGP) is as follows: The WGP reads the external file and interprets vectors with a file slice integer value of one as a scan vector. It assigns the first 5 pin codes of the file slice codes.all field to a temporary string. Next it merges the elements of the temporary string into their appropriate positions in the pin codes.all string. Finally, the apply procedure is called to apply the scan vector to only the scan pins. For parallel vectors (file slice integer greater than one), the processing consists of simply applying the vector.

The following is the sketch of the body of a waveform generation procedure for reading and applying vectors from the external file discussed in the previous section.

```
loop
  READ_FILE_SLICE( VECTOR_FILE, VECTOR );
  exit when VECTOR.END_OF_FILE;
  if VECTOR.FS_INTEGER = 1 then
    SCAN_STRING( 1 to 5 ) := VECTOR.CODES.all( 1 to 5 );
    VECTOR.CODES.all :=
        MERGE_STRING( VECTOR.CODES.all, SCAN_STRING( 1 ), TCK );
    VECTOR.CODES.all :=
        MERGE_STRING( VECTOR.CODES.all, SCAN_STRING( 2 ), TMS );
    VECTOR.CODES.all :=
        MERGE_STRING( VECTOR.CODES.all, SCAN_STRING( 3 ), TDI );
    VECTOR.CODES.all :=
        MERGE_STRING( VECTOR.CODES.all, SCAN_STRING( 4 ), TDO );
    VECTOR.CODES.all :=
       MERGE_STRING( VECTOR.CODES.all, SCAN_STRING( 5 ), TRST );
```

```
      APPLY( WPL, VECTOR.CODES.all,
                  TIMING_1, SCAN_PINS );
    else
      APPLY( WPL, VECTOR.CODES.all, TIMING_2);
    end if;
  end loop;
```

Here, VECTOR is the **FILE_SLICE** variable to read the file slices. **VECTOR_FILE** is the external data file containing the serial and parallel vectors. The loop statement is used to repeatedly read and apply the serial/parallel vectors until the end of the file is encountered. The WGP obtains a file slice and checks whether the file slice integer is one. If not, the vector is applied as a parallel vector. If it is one, then it is processed as a serial vector.

In the case of serial vectors, the WGP first stores the pin codes, in the vector just read, in a temporary string called **SCAN_STRING**. The next five statements are used to copy the pin codes for the scan pins into their correct positions in the pin code string, to be given as an argument to the APPLY procedure. The CODES.all field in VECTOR is used for this purpose.

Merge_string is a predefined function provided by the WAVES standard packages. This function accepts a string as the first parameter, a character as the second parameter and a test pin as the last parameter. It uses the test pin's position in the **Test_Pins** enumeration as an index into the string and replaces the character at that position in the string with the character passed as the second parameter. (There are several **Merge_string** functions provided by the WAVES standard; please see the WAVES language reference manual for a description of all the other variants.)

The APPLY in the *then* clause of the conditional statement applies the pin codes to the scan pins only. **SCAN_PINS** is a pinset that contains TCK, TMS, TDI, TDO and TRST. The APPLY in the *else* clause of the conditional statement applies the pin codes to all the pins (parallel vector).

TIMING_1 and **TIMING_2** are two **WAVE_TIMING** objects denoting delay times and time data for the serial and parallel vectors respectively.

12.4 Boundary-Scan Example

Consider an Octal D type flip-flop device which has scan capability. It has the following test pins:

```
type TEST_PINS is (
 TCK, TMS, TDI, TDO,
```

```
 CLK, OCB,
 D1, D2, D3, D4, D5, D6, D7, D8,
 Q1, Q2, Q3, Q4, Q5, Q6, Q7, Q8);
```

The following is an external file containing both parallel and serial vectors and written using our conventions:

```
0000000000000000000000 : 2 ;
0000101010101010101010 : 2 ;
0000010101010101010101 : 2 ;
0000000000000000000000 : 2 ; % Initial state for serial test.

% Serial Data Begin
% TCK, TMS, TDI, TDO
11X- : 1;
11X- : 1;
10X- : 1;

100- : 1;
101- : 1;
101- : 1;
101- : 1;
100- : 1;
100- : 1;
100- : 1;
101- : 1;

11X- : 1;
11X- : 1;
11X- : 1;
10X- : 1;

100- : 1;
100- : 1;

11X- : 1;
11X- : 1;
11X- : 1;
11X- : 1;
10X- : 1;

101- : 1;
100- : 1;
100- : 1;
101- : 1;
100- : 1;
```

```
100- : 1;
100- : 1;
100- : 1;

11X- : 1;
11X- : 1;
10X- : 1;

% Serial Data End

10000101010100000000 : 2 ;
10000010101000000000 : 2 ;
10000101010100000000 : 2 ;
10000000000000000000 : 2 ;
```

The following is a waveform generation procedure which reads and applies the above external data file.

```
procedure waveform (signal wpl: inout WAVES_PORT_LIST) is

      -- Declare the Scan string
      variable SCAN_STRING : string( 1 to 5 );


      -- Declare external file
      file VECTOR_FILE : text is in "vectors.txt";

      -- Declare file slice variable
      variable VECTOR : FILE_SLICE := NEW_FILE_SLICE;

      -- Scan pins pinset
      constant SCAN_PINS : PINSET := new_pinset((TCK, TMS, TDI, TDO));

 constant vector_FSA : Frame_Set_Array :=
  New_frame_set_array(Pulse_high(50 ns, 100ns), TCK) +
  New_frame_set_array(Non_return(0 ns), ALL_PINS AND NOT
new_pinset(TCK));

 variable timing: Time_Data := new_time_data(vector_FSA);

 variable pin_timing : WAVE_TIMING := (
   Delay(Etime(100 ns)), timing );

      begin -- waveform generator procedure

         loop
           READ_FILE_SLICE( VECTOR_FILE, VECTOR );
           exit when VECTOR.END_OF_FILE;
           if VECTOR.FS_INTEGER = 1 then
```

```
            SCAN_STRING( 1 to 5 ) := VECTOR.CODES.all( 1 to 5 );
            VECTOR.CODES.all :=
                MERGE_STRING( VECTOR.CODES.all, SCAN_STRING( 1 ), TCK );
            VECTOR.CODES.all :=
                MERGE_STRING( VECTOR.CODES.all, SCAN_STRING( 2 ), TMS );
            VECTOR.CODES.all :=
                MERGE_STRING( VECTOR.CODES.all, SCAN_STRING( 3 ), TDI );
            VECTOR.CODES.all :=
                MERGE_STRING( VECTOR.CODES.all, SCAN_STRING( 4 ), TDO );
            APPLY( WPL, VECTOR.CODES.all, pin_timing, SCAN_PINS );
          else
            APPLY( WPL, VECTOR.CODES.all, pin_timing,
                                 (ALL_PINS AND NOT SCAN_PINS);
          end if;
        end loop;

end waveform;
```

The waveform generation procedure is self-explanatory. Recall that **ALL_PINS** is a pinset that contains all the test pins and is a predefined constant in the WAVES standard.

In ***Summary of Chapter 12***, we have described how WAVES can be used to test devices with boundary-scan capability. We also illustrated how, by following suitable conventions, reusable packages of procedures can be developed for use in a set of related applications, boundary-scan architectures in this case. Such soft standards should be developed and used whenever possible to facilitate maximum reuse of WAVES/VHDL models.

12.5 References

1. "IEEE Standard Test Access Port and Boundary-Scan Architecture", IEEE Standard 1149.1-1990, IEEE Press, 1990.

2. C. M. Maunder and R. E. Tulloss, "The Test Access Port and Boundary Scan Architecture", IEEE Computer Society Press, 1990.

In Conclusion of our text on WAVES and VHDL application, we have attempted to provide material which will assist all users, from the novice to the seasoned practitioner. WAVES is truly a useful standard which can assist us in testing all the critical facets of our design, from simulation through actual hardware testing. We expect WAVES to ultimately become the popular and conventional means to test VHDL simulations, and to pace the evolution of VHDL as it evolves to suit the needs of designers. We hope the readers of this text will find useful guidance and assistance in their endeavors to incorporate WAVES into their design and testing environments.

APPENDICES

A - D

APPENDIX A

WAVES LOGIC VALUE SYSTEM FOR IEEE STD 1164-1993

```
library WAVES_STD;
use WAVES_STD.WAVES_Standard.all;
package WAVES_1164_Logic_Value is

   type Logic_value is
          (
           DONT_CARE,
           SENSE_X,
           SENSE_0,
           SENSE_1,
           SENSE_Z,
           SENSE_W,
           SENSE_L,
           SENSE_H,
           DRIVE_X,
           DRIVE_0,
           DRIVE_1,
           DRIVE_Z,
           DRIVE_W,
           DRIVE_L,
           DRIVE_H   );


  function Value_Dictionary( VALUE : Logic_value ) return
          Event_value;


end WAVES_1164_Logic_Value;


package body WAVES_1164_Logic_Value is

  function Value_Dictionary( VALUE : Logic_value ) return
          Event_value is

  begin
    case VALUE is
```

```
when DRIVE_X =>
  return state = UNKNOWN and
         strength = DRIVE and
         direction = STIMULUS;
when DRIVE_0 =>
  return state = LOW and
         strength = DRIVE and
         direction = STIMULUS;
when DRIVE_1 =>
  return state = HIGH and
         strength = DRIVE and
         direction = STIMULUS;
when DRIVE_Z =>
  return state = MIDBAND and
         strength = DISCONNECTED and
         direction = STIMULUS;
when DRIVE_W =>
  return state = UNKNOWN and
         strength = RESISTIVE and
         direction = STIMULUS;
when DRIVE_L =>
  return state = LOW and
         strength = RESISTIVE and
         direction = STIMULUS;
when DRIVE_H =>
  return state = HIGH and
         strength = RESISTIVE and
         direction = STIMULUS;
when SENSE_X =>
  return state = UNKNOWN and
         strength = DRIVE and
         direction = RESPONSE and
         relevance = REQUIRED;
when SENSE_0 =>
  return state = LOW and
         strength = DRIVE and
         direction = RESPONSE and
         relevance = REQUIRED;
when SENSE_1 =>
  return state = HIGH and
         strength = DRIVE and
         direction = RESPONSE and
         relevance = REQUIRED;
when SENSE_Z =>
  return state = MIDBAND and
         strength = DISCONNECTED and
         direction = RESPONSE and
```

```
        relevance = REQUIRED;
   when SENSE_W =>
    return state = UNKNOWN and
        strength = RESISTIVE and
        direction = RESPONSE and
        relevance = REQUIRED;
   when SENSE_L =>
    return state = LOW and
        strength = RESISTIVE and
        direction = RESPONSE and
        relevance = REQUIRED;
   when SENSE_H =>
    return state = HIGH and
        strength = RESISTIVE and
        direction = RESPONSE and
        relevance = REQUIRED;
   when DONT_CARE =>
    return UNSPECIFIED;
  end case;
 end Value_Dictionary;

end WAVES_1164_Logic_Value;
```

APPENDIX B

WAVES_1164_Pin_Codes

```
package WAVES_1164_Pin_Codes is

   constant Pin_codes : String := "X01ZWLH-";

end WAVES_1164_Pin_Codes;
```

APPENDIX C

WAVES_1164_Frames Package

```
Library WAVES_1164;
use WAVES_1164.WAVES_1164_Pin_Codes.all;
use WAVES_1164.WAVES_1164_Logic_Value.all;
use WAVES_1164.WAVES_Interface.all;
package WAVES_1164_Frames is

  --
  -- Declare functions that return Frame Sets.
  --

  function Non_Return( T1 : Time ) return Frame_set;

  function Return_Low( T1, T2 : Time ) return Frame_set;

  function Return_High( T1, T2 : Time ) return Frame_set;

  function Return_Complement( T1, T2 : Time ) return Frame_set;

  function Pulse_Low( T1, T2 : Time ) return Frame_set;

  function Pulse_Low_Skew( T0, T1, T2 : Time ) return Frame_set;

  function Pulse_High( T1, T2 : Time ) return Frame_set;

  function Pulse_High_Skew( T0, T1, T2 : Time ) return Frame_set;

  function Window( T1, T2 : Time ) return Frame_Set;

  function Window_Skew( T0, T1, T2 : Time ) return Frame_Set;

end WAVES_1164_Frames;

package body WAVES_1164_Frames is

  --
  -- Frame Set function definitions.
```

```
--

function Non_Return( T1 : Time ) return Frame_set is

  constant EDGE : Event_time := Etime( T1 );

begin
  return
    New_Frame_Set( 'X', Frame_Event( (DRIVE_X, EDGE) ) ) +
    New_Frame_Set( '0', Frame_Event( (DRIVE_0, EDGE) ) ) +
    New_Frame_Set( '1', Frame_Event( (DRIVE_1, EDGE) ) ) +
    New_Frame_Set( 'Z', Frame_Event( (DRIVE_Z, EDGE) ) ) +
    New_Frame_Set( 'W', Frame_Event( (DRIVE_W, EDGE) ) ) +
    New_Frame_Set( 'L', Frame_Event( (DRIVE_L, EDGE) ) ) +
    New_Frame_Set( 'H', Frame_Event( (DRIVE_H, EDGE) ) ) +
    New_Frame_Set( '-', Frame_Event );
end Non_Return;


function Return_Low( T1, T2 : Time ) return Frame_set is

  constant EDGE1 : Event_time := Etime( T1 );
  constant EDGE2 : Event_time := Etime( T2 );

begin
  assert T1 < T2
  report "Timing violation in Return_Low frames.  " &
        "The inequality : T1 < T2 Must hold."
  severity FAILURE;
  return
    New_Frame_Set( 'X', Frame_Elist( ((DRIVE_X, EDGE1),
                                (DRIVE_0, EDGE2)) ) ) +
    New_Frame_Set( '0', Frame_Event( (DRIVE_0, EDGE1) ) ) +
    New_Frame_Set( '1', Frame_Elist( ((DRIVE_1, EDGE1),
                                (DRIVE_0, EDGE2)) ) ) +
    New_Frame_Set( 'Z', Frame_Elist( ((DRIVE_Z, EDGE1),
                                (DRIVE_0, EDGE2)) ) ) +
    New_Frame_Set( 'W', Frame_Elist( ((DRIVE_W, EDGE1),
                                (DRIVE_0, EDGE2)) ) ) +
    New_Frame_Set( 'L', Frame_Elist( ((DRIVE_L, EDGE1),
                                (DRIVE_0, EDGE2)) ) ) +
    New_Frame_Set( 'H', Frame_Elist( ((DRIVE_H, EDGE1),
                                (DRIVE_0, EDGE2)) )) +
    New_Frame_Set( '-', Frame_Event( (DRIVE_0, EDGE2 ) ) );
end Return_Low;
```

```
function Return_High( T1, T2 : Time ) return Frame_set is

  constant EDGE1 : Event_time := Etime( T1 );
  constant EDGE2 : Event_time := Etime( T2 );

begin
  assert T1 < T2
  report "Timing violation in Return_High frames.  " &
       "The inequality: T1 < T2 Must hold."
  severity FAILURE;
  return
    New_Frame_Set( 'X', Frame_Elist( ((DRIVE_X, EDGE1),
                              (DRIVE_1, EDGE2)) ) ) +
    New_Frame_Set( '0', Frame_Elist( ((DRIVE_0, EDGE1),
                              (DRIVE_1, EDGE2)) ) ) +
    New_Frame_Set( '1', Frame_Event( (DRIVE_1, EDGE1) ) ) +
    New_Frame_Set( 'Z', Frame_Elist( ((DRIVE_Z, EDGE1),
                              (DRIVE_1, EDGE2)) ) ) +
    New_Frame_Set( 'W', Frame_Elist( ((DRIVE_W, EDGE1),
                              (DRIVE_1, EDGE2)) ) ) +
    New_Frame_Set( 'L', Frame_Elist( ((DRIVE_L, EDGE1),
                              (DRIVE_1, EDGE2)) ) ) +
    New_Frame_Set( 'H', Frame_Elist( ((DRIVE_H, EDGE1),
                              (DRIVE_1, EDGE2)) )) +
    New_Frame_Set( '-', Frame_Event( (DRIVE_1, EDGE2 ) ) );
end Return_High;


function Return_Complement( T1, T2 : Time ) return Frame_set is

  constant EDGE1 : Event_time := Etime( T1 );
  constant EDGE2 : Event_time := Etime( T2 );

begin
  assert T1 < T2
  report "Timing violation in Return_Complement frames.  " &
       "The inequality: T1 < T2 Must hold."
  severity FAILURE;
  return
    New_Frame_Set( 'X', Frame_Event( (DRIVE_X, EDGE1) ) ) +
    New_Frame_Set( '0', Frame_Elist( ((DRIVE_0, EDGE1),
                              (DRIVE_1, EDGE2)) ) ) +
    New_Frame_Set( '1', Frame_Elist( ((DRIVE_1, EDGE1),
                              (DRIVE_0, EDGE2)) ) ) +
    New_Frame_Set( 'Z', Frame_Event( (DRIVE_Z, EDGE1) ) ) +
    New_Frame_Set( 'W', Frame_Event( (DRIVE_W, EDGE1) ) ) +
    New_Frame_Set( 'L', Frame_Elist( ((DRIVE_L, EDGE1),
```

```
                              (DRIVE_H, EDGE2)) ) ) +
   New_Frame_Set( 'H', Frame_Elist( ((DRIVE_H, EDGE1),
                              (DRIVE_L, EDGE2)) )) +
   New_Frame_Set( '-', Frame_Event );
 end Return_Complement;


 function Pulse_Low( T1, T2 : Time ) return Frame_set is

  constant EDGE0 : Event_time := Etime( 0 ns );
  constant EDGE1 : Event_time := Etime( T1 );
  constant EDGE2 : Event_time := Etime( T2 );

 begin
  assert T1 < T2
  report "Timing violation in Pulse_Low frames.  " &
       "The inequality: T1 < T2 Must hold."
  severity FAILURE;
  return
   New_Frame_Set( 'X', Frame_Event ) +
   New_Frame_Set( '0', Frame_Elist( ((DRIVE_1, EDGE0),
                             (DRIVE_0, EDGE1),
                             (DRIVE_1, EDGE2)) ) ) +
   New_Frame_Set( '1', Frame_Event( (DRIVE_1, EDGE0) ) ) +
   New_Frame_Set( 'Z', Frame_Event ) +
   New_Frame_Set( 'W', Frame_Event ) +
   New_Frame_Set( 'L', Frame_Elist( ((DRIVE_H, EDGE0),
                             (DRIVE_L, EDGE1),
                             (DRIVE_H, EDGE2)) ) ) +
   New_Frame_Set( 'H', Frame_Event( (DRIVE_H, EDGE0) ) ) +
   New_Frame_Set( '-', Frame_Event );
end Pulse_Low;


function Pulse_Low_Skew( T0, T1, T2 : Time ) return Frame_set is

 constant EDGE0 : Event_time := Etime( T0 );
 constant EDGE1 : Event_time := Etime( T1 );
 constant EDGE2 : Event_time := Etime( T2 );

begin
 assert T0 < T1 and T1 < T2
 report "Timing violation in Pulse_Low frames.  " &
      "The inequality: T0 < T1 < T2 Must hold."
 severity FAILURE;
 return
  New_Frame_Set( 'X', Frame_Event ) +
```

```
    New_Frame_Set( '0', Frame_Elist( ((DRIVE_1, EDGE0),
                            (DRIVE_0, EDGE1),
                            (DRIVE_1, EDGE2)) ) ) +
    New_Frame_Set( '1', Frame_Event( (DRIVE_1, EDGE0) ) ) +
    New_Frame_Set( 'Z', Frame_Event ) +
    New_Frame_Set( 'W', Frame_Event ) +
    New_Frame_Set( 'L', Frame_Elist( ((DRIVE_H, EDGE0),
                            (DRIVE_L, EDGE1),
                            (DRIVE_H, EDGE2)) ) ) +
    New_Frame_Set( 'H', Frame_Event( (DRIVE_H, EDGE0) ) ) +
    New_Frame_Set( '-', Frame_Event );
end Pulse_Low_Skew;


function Pulse_High( T1, T2 : Time ) return Frame_set is

  constant EDGE0 : Event_time := Etime( 0 ns );
  constant EDGE1 : Event_time := Etime( T1 );
  constant EDGE2 : Event_time := Etime( T2 );

begin
  assert T1 < T2
  report "Timing violation in Pulse_High frames.  " &
      "The inequality: T1 < T2 Must hold."
  severity FAILURE;
  return
    New_Frame_Set( 'X', Frame_Event ) +
    New_Frame_Set( '0', Frame_Event( (DRIVE_0, EDGE0) ) ) +
    New_Frame_Set( '1', Frame_Elist( ((DRIVE_0, EDGE0),
                            (DRIVE_1, EDGE1),
                            (DRIVE_0, EDGE2)) ) ) +
    New_Frame_Set( 'Z', Frame_Event ) +
    New_Frame_Set( 'W', Frame_Event ) +
    New_Frame_Set( 'L', Frame_Event( (DRIVE_L, EDGE0) ) ) +
    New_Frame_Set( 'H', Frame_Elist( ((DRIVE_L, EDGE0),
                            (DRIVE_H, EDGE1),
                            (DRIVE_L, EDGE2)) ) ) +
    New_Frame_Set( '-', Frame_Event );
end Pulse_High;


function Pulse_High_Skew( T0, T1, T2 : Time ) return Frame_set is

  constant EDGE0 : Event_time := Etime( T0 );
  constant EDGE1 : Event_time := Etime( T1 );
  constant EDGE2 : Event_time := Etime( T2 );
```

```
begin
  assert T0 < T1 and T1 < T2
  report "Timing violation in Pulse_High frames.  " &
       "The inequality: T0 < T1 < T2 Must hold."
  severity FAILURE;
  return
    New_Frame_Set( 'X', Frame_Event ) +
    New_Frame_Set( '0', Frame_Event( (DRIVE_0, EDGE0) ) ) +
    New_Frame_Set( '1', Frame_Elist( ((DRIVE_0, EDGE0),
                          (DRIVE_1, EDGE1),
                          (DRIVE_0, EDGE2)) ) ) +
    New_Frame_Set( 'Z', Frame_Event ) +
    New_Frame_Set( 'W', Frame_Event ) +
    New_Frame_Set( 'L', Frame_Event( (DRIVE_L, EDGE0) ) ) +
    New_Frame_Set( 'H', Frame_Elist( ((DRIVE_L, EDGE0),
                          (DRIVE_H, EDGE1),
                          (DRIVE_L, EDGE2)) ) ) +
    New_Frame_Set( '-', Frame_Event );
end Pulse_High_Skew;


function Window( T1, T2 : Time ) return Frame_set is

  constant EDGE0 : Event_time := Etime( 0 ns );
  constant EDGE1 : Event_time := Etime( T1 );
  constant EDGE2 : Event_time := Etime( T2 );

begin
  assert T1 < T2
  report "Timing violation in Window frames.  " &
       "The inequality: T1 < T2 Must hold."
  severity FAILURE;
  return
    New_Frame_Set( 'X', Frame_Elist( ((DONT_CARE, EDGE0),
                          (SENSE_X, EDGE1),
                          (DONT_CARE, EDGE2)) ) ) +
    New_Frame_Set( '0', Frame_Elist( ((DONT_CARE, EDGE0),
                          (SENSE_0, EDGE1),
                          (DONT_CARE, EDGE2)) ) ) +
    New_Frame_Set( '1', Frame_Elist( ((DONT_CARE, EDGE0),
                          (SENSE_1, EDGE1),
                          (DONT_CARE, EDGE2)) ) ) +
    New_Frame_Set( 'Z', Frame_Elist( ((DONT_CARE, EDGE0),
                          (SENSE_Z, EDGE1),
                          (DONT_CARE, EDGE2)) ) ) +
    New_Frame_Set( 'W', Frame_Elist( ((DONT_CARE, EDGE0),
                          (SENSE_W, EDGE1),
```

```
                            (DONT_CARE, EDGE2)) ) ) +
    New_Frame_Set( 'L', Frame_Elist( ((DONT_CARE, EDGE0),
                            (SENSE_L, EDGE1),
                            (DONT_CARE, EDGE2)) ) ) +
    New_Frame_Set( 'H', Frame_Elist( ((DONT_CARE, EDGE0),
                            (SENSE_H, EDGE1),
                            (DONT_CARE, EDGE2)) ) ) +
    New_Frame_Set( '-', Frame_Event( (DONT_CARE, EDGE0) ));
 end Window;

 function Window_Skew( T0, T1, T2 : Time ) return Frame_set is

  constant EDGE0 : Event_time := Etime( T0 );
  constant EDGE1 : Event_time := Etime( T1 );
  constant EDGE2 : Event_time := Etime( T2 );

 begin
  assert T0 < T1 and T1 < T2
  report "Timing violation in Window frames. " &
       "The inequality: T0 < T1 < T2 Must hold."
  severity FAILURE;
  return
    New_Frame_Set( 'X', Frame_Elist( ((DONT_CARE, EDGE0),
                            (SENSE_X, EDGE1),
                            (DONT_CARE, EDGE2)) ) ) +
    New_Frame_Set( '0', Frame_Elist( ((DONT_CARE, EDGE0),
                            (SENSE_0, EDGE1),
                            (DONT_CARE, EDGE2)) ) ) +
    New_Frame_Set( '1', Frame_Elist( ((DONT_CARE, EDGE0),
                            (SENSE_1, EDGE1),
                            (DONT_CARE, EDGE2)) ) ) +
    New_Frame_Set( 'Z', Frame_Elist( ((DONT_CARE, EDGE0),
                            (SENSE_Z, EDGE1),
                            (DONT_CARE, EDGE2)) ) ) +
    New_Frame_Set( 'W', Frame_Elist( ((DONT_CARE, EDGE0),
                            (SENSE_W, EDGE1),
                            (DONT_CARE, EDGE2)) ) ) +
    New_Frame_Set( 'L', Frame_Elist( ((DONT_CARE, EDGE0),
                            (SENSE_L, EDGE1),
                            (DONT_CARE, EDGE2)) ) ) +
    New_Frame_Set( 'H', Frame_Elist( ((DONT_CARE, EDGE0),
                            (SENSE_H, EDGE1),
                            (DONT_CARE, EDGE2)) ) ) +
    New_Frame_Set( '-', Frame_Event( (DONT_CARE, EDGE0) ));
 end Window_Skew;

end WAVES_1164_Frames;
```

APPENDIX D

WAVES_1164_Utilities Package

```
library waves_std;
use waves_std.waves_system.all;

library IEEE;
use IEEE.STD_Logic_1164.all;

package WAVES_1164_utilities is

----------------------------------------
-- Procedure and Function Declarations --
----------------------------------------
--

 -- This function converts a waves port element to 1164 std_logic bit.
 -- The Translation is specific to the order of logic value definitions
 -- defined in the waves_1164_logic_values package
 --
 FUNCTION stim_1164( Port_element : system_waves_port) RETURN std_logic;


 -- This function converts a waves port element to 1164 std_logic bits
 -- The Translation is specific to the order of logic value definitions
 -- defined in the waves_1164_logic_values package
 --
 FUNCTION stim_1164( Port_list : system_waves_port_list)
                    RETURN std_ulogic_vector;


 -- This function converts a waves port list to 1164 std_logic values
 -- The Translation is specific to the order of logic value definitions
 -- defined in the waves_1164_logic_values package
 --
 FUNCTION stim_1164( Port_list : system_waves_port_list)
                    RETURN std_logic_vector;


 -- This function converts a waves port element to 1164 std_logic bits
 -- IT ALSO strips off the drive values and replaces them with '-'
```

```
-- The Translation is specific to the order of logic value definitions
-- defined in the waves_1164_logic_values package
--
FUNCTION expect_1164( Port_element : system_waves_port) RETURN std_ulogic;


-- This function converts a waves port list to 1164 std_logic_vectors
-- IT ALSO strips off the drive values and replaces them with '-'
-- The Translation is specific to the order of logic value definitions
-- defined in the waves_1164_logic_values package
--
FUNCTION expect_1164( Port_list : system_waves_port_list)
                RETURN std_ulogic_vector;


-- This function converts a waves port element to 1164 std_logic bits
-- IT ALSO strips off the sense values and replaces them with 'Z'
-- The Translation is specific to the order of logic value definitions
-- defined in the waves_1164_logic_values package
--
FUNCTION bi_dir_1164( Port_element : system_waves_port) RETURN std_logic;


-- This function converts a waves port list to 1164 std_logic_vectors
-- IT ALSO strips off the sense values and replaces them with 'Z'
-- The Translation is specific to the order of logic value definitions
-- defined in the waves_1164_logic_values package
--
FUNCTION bi_dir_1164( Port_list : system_waves_port_list)
                RETURN std_logic_vector;


-- This function evaluates two 1164 std_logic bits for compatibility
-- The actual data bit is evaluated to determine if it is compatible
-- to a expected or predicted value.  The order must be preserved.
-- example if actual data is '-' and expected data '1' result = false
--         if actual data is '1' and expected data '-' result = true
--
FUNCTION compatible( actual: std_logic; expected : std_ulogic )
   RETURN BOOLEAN;


-- This function evaluates two 1164 std_logic_vectors for compatibility
-- The actual data bit is evaluated to determine if it is compatible
-- to a expected or predicted value.  The order must be preserved.
-- example if actual data is '-' and expected data '1' result = false
--         if actual data is '1' and expected data '-' result = true
```

```
--
FUNCTION compatible( actual: std_ulogic_vector; expected : std_ulogic_vector)
   RETURN BOOLEAN;
-- This function evaluates two 1164 std_logic_vectors for compatibility
-- The actual data bit is evaluated to determine if it is compatible
-- to a expected or predicted value.  The order must be preserved.
-- example if actual data is '-' and expected data '1' result = false
--        if actual data is '1' and expected data '-' result = true
--
FUNCTION compatible( actual: std_logic_vector; expected : std_ulogic_vector)
   RETURN BOOLEAN;


end WAVES_1164_utilities;

package body WAVES_1164_utilities is

------------------------------
-- Internal Type Definitions --
------------------------------
--
 type Boolean_matrix is array ( STD_ulogic, STD_ulogic ) of Boolean;
 -- Define Compatible Table.   (Actual, Expected)
 --
 constant Compatible_Table : Boolean_matrix :=
 -- EXPECTED  EXPECTED  EXPECTED  EXPECTED  EXPECTED  EXPECTED
 -- U    X    0    1    Z    W    L    H    -
(( TRUE,  FALSE, FALSE, FALSE, FALSE, FALSE, FALSE, FALSE, TRUE ),  -- 'U'
 ( FALSE, TRUE,  FALSE, FALSE, FALSE, FALSE, FALSE, FALSE, TRUE ),  -- 'X'  A
 ( FALSE, TRUE,  TRUE,  FALSE, FALSE, FALSE, FALSE, FALSE, TRUE ),  -- '0'  C
 ( FALSE, TRUE,  FALSE, TRUE,  FALSE, FALSE, FALSE, FALSE, TRUE ),  -- '1'  T
 ( FALSE, FALSE, FALSE, FALSE, TRUE,  FALSE, FALSE, FALSE, TRUE ),  -- 'Z'  U
 ( FALSE, FALSE, FALSE, FALSE, FALSE, TRUE,  FALSE, FALSE, TRUE ),  -- 'W'  A
 ( FALSE, FALSE, FALSE, FALSE, FALSE, TRUE,  TRUE,  FALSE, TRUE ),  -- 'L'  L
 ( FALSE, FALSE, FALSE, FALSE, FALSE, TRUE,  FALSE, TRUE,  TRUE ),  -- 'H'
 ( FALSE, FALSE, FALSE, FALSE, FALSE, FALSE, FALSE, FALSE, TRUE )); -- '-'
--

 type sim_code_array is array (Natural range <> ) of std_ulogic;
 constant Translate : Sim_code_array := ('-','X','0','1','Z','W','L','H',
                                'X','0','1','Z','W','L','H');
                              -- 0  1  2  3  4  5  6  7
                              --    8  9  10 11 12 13 14

--
---------------------------
-- Procedure and Function --
```

```
---------------------------

-- This function converts a waves port element to 1164 std_logic bits
-- The Translation is specific to the order of logic value definitions
-- defined in the waves_1164_logic_values package
--
FUNCTION stim_1164( Port_element : system_waves_port)
                        RETURN std_logic IS
BEGIN
  RETURN Translate( Port_element.l_value );
END stim_1164;


-- This function converts a waves port list to 1164 std_logic_vectors
-- The Translation is specific to the order of logic value definitions
-- defined in the waves_1164_logic_values package
--
FUNCTION stim_1164( Port_list : system_waves_port_list)
                RETURN std_ulogic_vector is

 VARIABLE r : std_ulogic_vector(port_list'range);
  BEGIN
  For i IN port_list'range LOOP
   r(i):= Translate( port_list(i).l_value );
  END LOOP;
  RETURN r;
END stim_1164;


-- This function converts a waves port list to 1164 std_logic_vectors
-- The Translation is specific to the order of logic value definitions
-- defined in the waves_1164_logic_values package
--
FUNCTION stim_1164( Port_list : system_waves_port_list)
                RETURN std_logic_vector is

 VARIABLE r : std_logic_vector(port_list'range);
  BEGIN
  For i IN port_list'range LOOP
   r(i):= Translate( port_list(i).l_value );
  END LOOP;
  RETURN r;
END stim_1164;


-- This function converts a waves port element to 1164 std_logic bits
-- IT ALSO strips off the drive values and replaces them with '-'
```

```
-- The Translation is specific to the order of logic value definitions
-- defined in the waves_1164_logic_values package
--
FUNCTION expect_1164( Port_element : system_waves_port)
                         RETURN std_ulogic IS
 VARIABLE Result : std_ulogic;
BEGIN
  IF (Port_element.l_value < 8 ) THEN
    Result:= Translate( Port_element.l_value );
  ELSE
    Result:='-';
  END IF;
  RETURN Result;
END expect_1164;


-- This function converts a waves port list to 1164 std_logic_vectors
-- IT ALSO strips off the drive values and replaces them with '-'
-- The Translation is specific to the order of logic value definitions
-- defined in the waves_1164_logic_values package
--
FUNCTION expect_1164( Port_list : system_waves_port_list)
                        RETURN std_ulogic_vector is

  VARIABLE r : std_ulogic_vector(port_list'range);
  BEGIN
   For i IN port_list'range LOOP
     IF (Port_list( i ).l_value < 8 ) THEN
       r(i):= Translate( port_list(i).l_value );
     ELSE
       r(i):='-';
     END IF;
   END LOOP;
   RETURN r;
  END expect_1164;


-- This function converts a waves port element to 1164 std_logic bits
-- IT ALSO strips off the sense values and replaces them with 'Z'
-- The Translation is specific to the order of logic value definitions
-- defined in the waves_1164_logic_values package
--
FUNCTION bi_dir_1164( Port_element : system_waves_port)
                         RETURN std_logic IS
 VARIABLE Result : std_logic;
BEGIN
  IF (Port_element.l_value > 7) THEN
```

```
    Result:= Translate( Port_element.l_value );
   ELSE
    Result:='Z';
   END IF;
   RETURN Result;
 END bi_dir_1164;


-- This function converts a waves port list to 1164 std_logic_vectors
-- IT ALSO strips off the sense values and replaces them with 'Z'
-- The Translation is specific to the order of logic value definitions
-- defined in the waves_1164_logic_values package
--
FUNCTION bi_dir_1164( Port_list : system_waves_port_list)
                        RETURN std_logic_vector is

  VARIABLE r : std_logic_vector(port_list'range);
  BEGIN
   For i IN port_list'range LOOP
    IF (Port_list( i ).l_value > 7 ) THEN
      r(i):= Translate( port_list(i).l_value );
    ELSE
      r(i):='Z';
    END IF;
   END LOOP;
   RETURN r;
END bi_dir_1164;


-- This function evaluates two 1164 std_logic bits for compatibility
-- The actual data bit is evaluated to determine if it is compatible
-- to a expected or predicted value.  The order must be preserved.
--
FUNCTION compatible ( actual : std_logic;
               expected : std_ulogic ) RETURN BOOLEAN IS
BEGIN
    RETURN ( compatible_table(actual, expected) );
END compatible;


-- This function evaluates two 1164 std_logic vectors for compatibility
-- The actual data bit is evaluated to determine if it is compatible
-- to a expected or predicted value.  The order must be preserved.
--
FUNCTION compatible (  actual : std_ulogic_vector;
               expected : std_ulogic_vector) RETURN BOOLEAN IS
  alias a :  STD_ulogic_vector (actual'length-1 DOWNTO 0) is actual;
```

```
  alias e :  STD_ulogic_vector (expected'length-1 DOWNTO 0) is expected;
 BEGIN
  assert e'length = a'length
    report "Vector Length's Incompatable in Compatible Function"
    severity failure;
   For Index in (expected'length -1) downto 0 loop
     IF not compatible_table( a(index), e(index)) THEN
       RETURN FALSE;
     END IF;
   END LOOP;
   RETURN TRUE;
 END compatible;


 -- This function evaluates two 1164 std_logic vectors for compatibility
 -- The actual data bit is evaluated to determine if it is compatible
 -- to a expected or predicted value.  The order must be preserved.
 --
 FUNCTION compatible (  actual : std_logic_vector;
                 expected : std_ulogic_vector) RETURN BOOLEAN IS
  alias a :  STD_logic_vector (actual'length-1 DOWNTO 0) is actual;
  alias e :  STD_ulogic_vector (expected'length-1 DOWNTO 0) is expected;
 BEGIN
  assert e'length = a'length
    report "Vector Length's Incompatable in Compatible Function"
    severity failure;
   For Index in (expected'length -1) downto 0 loop
     IF not compatible_table( a(index), e(index)) THEN
       RETURN FALSE;
     END IF;
   END LOOP;
   RETURN TRUE;
 END compatible;

END WAVES_1164_utilities;
```

INDICES

Application Index

Topic Index